Creation &
the Environment

CENTER BOOKS IN ANABAPTIST STUDIES

Donald B. Kraybill, *Consulting Editor*
George F. Thompson, *Series Founder and Director*

Published in cooperation with the Center for American Places,
Santa Fe, New Mexico, and Harrisonburg, Virginia

Creation &

The Johns Hopkins University Press

Baltimore & London

the Environment

An Anabaptist Perspective
on a Sustainable World

Edited by

CALVIN REDEKOP

This book was brought to publication with the generous assistance of the
Environmental Task Force of the Mennonite Church.

The Johns Hopkins University Press
2715 North Charles Street
Baltimore, Maryland 21218-4363
www.press.jhu.edu

Library of Congress Cataloging-in-Publication Data

Creation and the environment : an Anabaptist perspective on
a sustainable world / edited by Calvin Redekop.
p.   cm. — (Center books in Anabaptist studies)
Includes bibliographical references and index.
ISBN 0-8018-6422-4 (alk. paper)—ISBN 0-8018-6423-2 (pbk. : alk. paper).
1. Human ecology—Religious aspects—Anabaptists.   2. Human ecology—
Religious aspects—Mennonites.   I. Redekop, Calvin Wall, 1925–
II. Series.
BX4931.2C.74 2000
261.8'362'088243—DC21   00-008081
A catalog record for this book is available from the British Library.

Engravings by Jan Luyken (1649–1712)
Frontispiece: In the Beginning . . . Genesis 1:1–2. *Part I:* The Bountiful Garden and Adam and
Eve's Disobedience . . . Genesis 3:13. *Part II:* Noah Preserves All Living Things in the Ark . . .
Genesis 7:1–5. *Part III:* Elijah Fed by the Ravens . . . 1 Kings 17:6.
Part IV: The Tree of Life . . . Revelation 22:2.

To the increasing numbers of persons of all traditions, including the Anabaptists, who have committed and are committing themselves to preserve the creation. May their tribe increase, as it must if creation is to be preserved.

Contents

PART III

Anabaptists' Theological & Historical Orientation

PART IV

The Challenge to Take Care of the Earth

Acknowledgments

In 1989, an Environmental Task Force (ETF) was established by a joint meeting of the Mennonite Church and the General Conference Mennonite Church to promote environmental issues. In 1995, the ETF organized a "Creation Summit," a conference to address environmental and theological concerns from an Anabaptist perspective. Some of the papers presented at this conference are included in this book, which represents a starting point for more discussion about our relationship with God and his creation.

I thank the ETF and the authors for their contributions in producing the substantive message of this work; I thank James Fairfield for helpful editing of an early draft of the manuscript and Jeremy Yoder for his expert reproductions of the classic Jan Luyken prints.

Special thanks go to George F. Thompson, president of the Center for American Places, and his colleagues at the Johns Hopkins University Press, especially Linda Forlifer, for believing in the message of this book and for their assistance in bringing it to publication.

Introduction

One hot summer afternoon in 1933, I was helping my parents gather some so-called hay for the cows. (It was composed of thistles and other weeds, since grass would not grow in such drought conditions.) My father suddenly stopped and pointed to a low-lying cloud. "Calvin," he said, "do you see that dark cloud stretching along the western horizon? That is a major dust storm, and it will be here in a short while. We'd better get home right now and take care of the livestock."

The dark cloud rose higher in the sky, and the sand- and debris-filled storm finally arrived. Its terrible fury remains vivid in my memory. The howling wind scattered the chickens that we had been unable to chase into the hen house. Tumbleweeds (Russian thistles) came rolling in from the southwest like an invading army and cartwheeled across the yard. The screaming wind tore doors from their hinges and blew sand into our hair, eyes, ears, and noses; we could hardly breathe. We huddled inside, with the sand settling everywhere. We felt that at any moment the storm might blow away the small prairie shack.

The wind raged for several days. After the dust storm had finally stopped, everything we ate had sand in it. What I remember most about that dust storm, however, is the almost total darkness, around noon. There seemed to be something unnatural about it getting dark in the middle of the day without an eclipse.

That year there was a total crop failure in Montana. In fact, almost every year until the early 1940s was a drought year throughout the Great Plains and the Midwest, with plagues of grasshoppers and dust storms visiting us, sometimes separately and sometimes together. The drought continued in spite of desperate prayer meetings in church pastored by my grandfather, where we implored God to "open the gates of heaven" with rain. The fervent prayers are as memorable as the storms. But the rains did not come. I remember Dad searching for the farm machines each spring; parked along the fence in the yard behind the machine shed, they had been totally covered by drifting sand.

So the farmers, having homesteaded there only about fifteen years earlier (in 1915),[1] began to move away, mainly to the West Coast, "where everything is green and it rains." My parents finally gave up, had an auction, and in 1937 moved the family to Oregon to begin life anew on a little farm. I was very sad that we were leaving, for by then I was a boy of thirteen years and loved the unsettled and open frontier life in Montana, with the coyotes howling at night, the unexplored hills and coulees, the branding of the new calves in the spring, the breaking of wild horses, and all that went with farming and ranching on the prairie. Slowly I accepted the conclusion everyone accepted—that our experience in Montana, and in the other states of the Dust Bowl, had been God's punishment for disobedience. Being relatively naive, I did not understand what that disobedience was all about.[2]

Many years later I began to realize, slowly and almost imperceptibly, that the Dust Bowl was not God's punishment for our spiritual misdeeds; rather, it was the result of terrible sins committed by my parents' generation. They had not only benefited by taking free land from the dispossessed Assiniboine and Sioux Indians but also, with plows and tractors, destroyed forever the terribly sensitive several-inch membrane of plant and animal life that had maintained the prairies for thousands of years. And, with that realization, I became an environmentalist forever.[3]

Fortunately, our common human family has managed to survive until the end of the millennium just passed, but many people doubt whether it can exist indefinitely unless we change our treatment of this finite earth, which sustains all living things. This fear is not a delusion of fainthearted or pessimistic people. It is supported by the reality that some civilizations have already died; these stories should caution those who say it cannot happen. Our records of the deaths of societies indicate that it could happen again, this time on a global scale—these dead societies are like that dark cloud in the Montana sky, warning us that a disastrous storm is approaching.

In a fascinating but sobering account of the demise of the exotic Polynesian society on Easter Island, Jared Diamond suggests that,

> among all vanquished civilizations, that of the former Polynesian society on Easter Island remains unsurpassed in mystery and isolation, especially from the island's gigantic stone statues and its impoverished landscape. In just a few centuries, the people of Easter Island wiped out their forests, drove their plants and animals to extinction, and saw their complex society spiral into chaos and cannibalism . . . Everyone who has seen the abandoned buildings of the Khmer, the Maya, or the Anasazi is immediately moved to ask the same question: Why did the societies that erected those structures disappear?[4]

Diamond concludes that the Easter Island civilization died "not with a bang, but with a whimper . . . The forest the islanders depended on didn't simply disappear one day—it vanished slowly, over decades . . . Any islander who tried to warn about the dangers of progressive deforestation would have been overridden by vested interests of carvers, bureaucrats, and chiefs, whose jobs depended on continued deforestation."[5] There is something very contemporary, relevant, and ominous about this description. If it could happen to an almost totally nonindustrialized society, what lies in store for members of a highly industrialized civilization?

Every human generation, from earliest times it seems, has tried to understand itself and its relationship to the natural and supernatural reality. But complicating that search are the systems of self-deception humans have created for themselves, resulting in nonrational analysis and even destructive solutions. One of the most serious self-deceptions is denial. Each of us has heard of people who refused to see a physician, fearing the results of the examination or diagnosis. We avoid bad news as long as possible. Assuming that life will continue as usual is a comforting position. Easter Island is a classic example of this syndrome.

Human societies have historically confronted crises, and they have often faced them courageously. Examples are the bloody turmoil of the Reformation and the ensuing century of religious wars and the cataclysmic spasms of the great world wars of the twentieth century, including the threat of annihilation by the atom bomb. But these crises,

monumental as they were, did not end in the demise of civilizations because humans saw the abyss and stepped back. The environmental apoc-
alypse is much more probable because, for most people, there is no clearly
visible abyss—the ominous Easter Island parable of death by impercep-
tible degrees is largely ignored.[6]

Earlier calamities were avoided because they were massive and critical, demanding immediate and decisive actions. The environmental crisis is different. As the example of Easter Island demonstrates, the environmental crisis steals in quietly at night and ends, not with a bang, but
with a whimper.

We may have inadequate empirical proof for the factors directly responsible for the destruction of the twenty-odd civilizations described
by Arnold Toynbee some decades ago.[7] Nevertheless, it is becoming
clear that many civilizations died because of drastic disjunctures from
their environments. They misused and ultimately destroyed the environment so it could no longer sustain human habitation, but this destruction unfolded very slowly, with the end coming so subtly that no
one noticed that the situation was becoming critical.

Is there an environmental crisis in our time? Many people would respond with a resounding No! These would be the optimistic skeptics, the
so-called cornucopians,[8] who have long denied the existence of a crisis,
likening the "panic" of the environmentalists to the fable of timid
Chicken Little. Is it not true that all those who tell the parable prove
thereby that the sky has not yet fallen and, hence, never will?[9]

Others, who believe that Earth is being threatened but are hesitant
to change their lifestyle, have retreated to a reassuring compromise by
invoking the Gaia principle, which proposes that the planet has a self-
preserving and self-limiting mechanism allowing it to adjust and survive. In this view, the environment may change dramatically and not be
what most humans would enjoy or desire, but it will adjust and survive,
as will humankind, which has shown an amazing ability to adapt to new
conditions. The problem with this view is that the resultant environment
may be spiritually, culturally, and physically a very undesirable habitat.

The contributing authors and I align ourselves with those who believe that, if something is not done soon, the sky will indeed fall. We
agree with Barry Commoner's warning: "We have come to a turning

point in the human habitation of the earth. Continued plundering and pollution of the earth, if unchecked, will eventually destroy the fitness of this planet as a place for human life."[10] Others believe that we have already gone beyond the point of no return.

We are environmentalists because of the signs around us, but our perspective and actions are not derived from a sudden panic about saving the world from destruction. Rather, we have inherited the Judeo-Christian world-view that human beings are given the responsibility to "tend the garden," to nurture creation as described in the first chapters of Genesis. Thus, our concern about God's creation, although obviously supported by present critical aspects, has an eternal and enduring time frame.

The fact is that the Judeo-Christian record regarding the tending of God's creation is dismal, which spurred the development of this book. The commandments in the Old and New Testament about stewardship and creation have been interpreted and rationalized in many self-serving ways. We propose that the Anabaptist/Mennonite tradition has taken a substantially different point of view, at least in theory, about how the Old and New Testament should be read regarding the environment.

Fortunately, the signally important original Judeo-Christian teaching that humans are commanded to care for the creation is now being supported by contemporary scientific enterprise, which, with increasing urgency, is providing empirical proof that humans are changing the world with serious negative consequences. Equally important is the growing realization among the rank and file that human beings are indeed altering and corroding the nature of our environment, a conviction that has been long denied. The question thus becomes, In what ways is what we are doing good or bad for the environment? Furthermore, Why should we care about whether what we are doing is good or bad?

One answer derives from our value system, which mandates how the creation should be treated. The Anabaptist/Mennonite tradition has developed a unique, though not totally consistent, philosophical and ethical position regarding the creation. Although this position is a derivative of the Judeo-Christian tradition, it has followed a different path regarding our relationship to the land, though it has not always been true to its tradition.

This book approaches the consequences of our impact on the environment from a very practical perspective—if we care about the coming generations and want to bequeath the world to them with a cultural system that supports sustainable life, then we must act now. This book is not a complete guide to achieving a sustainable world, but we are convinced that only an environmental ethic ultimately allowing for an indefinite coexistence of a biodiverse ecosystem is a satisfactory solution. This book is relevant to that objective because the Anabaptist perspective offers some important insights toward achieving sustainability.[11]

The contributors to this book share the conviction that Anabaptism is based on a world-view that is significant for the health of the environment. We do not claim that Anabaptists have an exclusive insight into the ultimate purpose of the creation and how we are to relate to it. Nor do we deny that Anabaptism has been, and continues at times to be, destructive of the environment. But we present the Anabaptist/Mennonite tradition as a positive contribution, on balance, to the search for sustainable biodiversity.[12]

David Kline's engaging account of the Amish views of creation (chap. 4) and Lawrence Hart's winsome and inspiring Cheyenne perspective on our relationship to our mother earth (chap. 11) suggest that several traditions, as different as those of Anabaptists and American Indians, converge at many important points regarding sustainability. Hart believes that Anabaptism is sufficiently sensitive to the earth to allow him as a Cheyenne to become a member of the Anabaptist family. Who knows, there may yet be a song all humans can sing as we thank the Great Spirit for the bountiful harvest potentially available for all for ages to come—a song that Native Americans, Anabaptists, and all humankind will have to sing if an inhabitable world is to survive, countering the threats to creation rampant in our self-gratification-oriented, consumptive epoch.[13]

We are hopeful that this book will help citizens throughout the world to achieve a new approach to understanding the environment and how they may contribute to saving it. Readers may appreciate more fully our nurturing mother earth and may recognize our complicity in her fate and our future.

We offer these pages in the spirit of Chief Seattle, who is believed to have said: "This we know: The earth does not belong to man, man belongs to the earth. This we know: All things are connected like the blood which unites our family. All things are connected. Whatever befalls the earth befalls the sons of the earth. Man did not weave the web of life, he is merely a strand in it. Whatever he does to the web, he does to himself."[14]

PART I

Human Activities & Their
Alteration of the Creation

CHAPTER 1

Economics, Development, and Creation

James M. Harder and Karen Klassen Harder

TWO VISIONS FOR THE FUTURE

The concluding years of the twentieth century were marked by an ongoing struggle between two competing economic visions for the future. At the center of the struggle are profoundly different perspectives on economic growth and on its relationship to the twin problems of global poverty and environmental decay.

One vision for the future views economic growth—the increase over time of produced goods and services—as fundamental to the solution of nearly all significant social problems facing humanity. Steadily growing economic output is seen as essential for creating jobs, raising living standards among the world's poor, funding technological research, affording senior care, setting aside nature preserves, building better schools, producing more food, and freeing up more time for leisurely pursuits. Progress is assumed to go hand in hand with rising consumerism. Most simply put, there can never be too much economic growth.

An emerging alternative economic vision suggests that the drive for unrestrained economic growth itself has become the most important problem facing humanity. The primary concern is that a world of over six billion people striving for material satisfaction is drawing ever more heavily from finite supplies of natural resources to fuel an economic growth model destined to lead to an ecological disaster and global poverty without precedence.[1] There is an urgent need to find an economic way of life that is environmentally sustainable.

There could hardly exist a sharper divide between the two conflicting viewpoints on economic growth. Are current growth-oriented economic systems and institutions, in both their capitalistic and socialistic manifestations, creating a world of greater promise or greater peril for humanity and the rest of creation? From environmental and social standpoints, are these systems even sustainable? These immensely important questions are far from the proprietary concern of either economists or biologists—they involve the political process at every level and scholars from nearly all academic disciplines. Increasingly, the far-reaching implications of this debate are pushing it into the religious realm as well.

Among those who look to a bright future under present growth-oriented economic institutions and policies are the champions of the free market, the free trade economic doctrine that reigns supreme in the post–cold war world order. This group, led by the International Monetary Fund, the World Bank, the World Trade Organization, and the governments of leading industrialized nations, envisions continuous economic growth and material prosperity propelled by unimpeded global competition and free access to global markets. Their operational logic reflects their sense of duty to develop latent resources into their actualized state for the benefit of a global population that can never have enough. Their policies strive for ever-greater production efficiencies and freedom for low-cost producers to supply ever-expanding markets. With few caveats, these economic policies willingly entrust the future of society and all of creation to market forces of supply and demand.

As proof of their policy wisdom, this group offers its track record of progress: During the last quarter century alone, the world's economic output has nearly doubled, with several of the lower-income nations—including such newly industrialized countries as South Korea, Taiwan, Hong Kong, and Singapore—experiencing some of the most rapid rates of economic growth. During this same period, the already enormous American economy enlarged by 65 percent.[2] Economic growth advocates readily acknowledge that critical social and economic problems remain unsolved but are confident that still more growth is the surest route to their eradication. The authoritative *World Development Report* published by the World Bank assumes that economic growth must continue, and indeed accelerate, with world output increasing 3.5 times by the year 2030.[3] Summarized one popular news magazine, "Never have so many people been getting better-off so fast."[4]

The alternative perspective describes a much different picture of reality. It is also a less focused picture, coming from a multitude of disconnected voices around the world, mostly outside official circles. The message is complex and multifaceted, reflecting a real world of differing contexts. But it solidifies around one general theme: The growth of the global market economy is not delivering what it advertises—broadly shared prosperity and the prospect of a happy, secure future. Suggests author David Korten, "We are experiencing accelerating social and environmental disintegration in nearly every country of the world."[5]

The evidence for this point of view is mounting. From the social perspective, there is no denying that economic growth has spread its benefits very unevenly. In nearly every country the rich are getting richer while the poor stagnate with little real hope for advancement. Society is splitting farther apart along wealth lines, both between and within countries.

From an environmental perspective, the unrelenting push for economic development since the industrial age has exacted some very high costs from nature. Biologists warn of pushing the earth past the limits of its life-sustaining capacities. Since 1972, the world has lost 200 million hectares of trees (an area equivalent to all of the United States east of the Mississippi River), and each year the world's farmers lose an additional 25 billion tons of topsoil (the equivalent of all the wheat fields in Australia).[6] Air in the cities of some developing countries is now so polluted as to be considered life threatening, while in the United States, industry disposes of more than 10 billion pounds of toxins annually. In every country of the world (but especially among the world's richer countries), oil-fueled economic growth is sowing the seeds of serious long-term global warming due to increased levels of hydrocarbon emissions from cars, homes, and factories.[7]

Growth optimists argue that economic growth itself will eventually create both the technology and the financial capacity to eliminate any environmental and social problems due to the growth process. Growth pessimists are passionately skeptical of such claims. They see the depletion of the natural resource base as an ominous threat to both present and future generations.

FACTS, VALUES, AND MYTHS

Upon reflection, one thing quickly becomes apparent: The growth debate is not entirely a debate over facts. Greater precision of numbers, more reliable data, better economic and biological models are, of course, helpful. But inevitably the debate moves in and out of the realm of values fundamental to each individual's sense of self and the world. These values are not easily discussed or changed. They include one's personal understanding of ultimate life objectives for both self and society and the ranking of these objectives. They include one's understanding of the position, rights, and responsibilities of humans within the rest of the created order. They include one's understanding of the principles of fairness, justice, and responsibility for the welfare of others, now and in future generations.

The inevitable interplay of facts and values in any discussion of economic policy is guaranteed to yield differences of opinion. That is, after all, why a common joke holds that, if all economists in the world were laid end to end, they still wouldn't reach a consensus! Still, it is at least possible to sharpen the arguments, which we do here by exposing three powerful myths surrounding economic growth.

Myth 1: Growth Equals Development

The word *growth* is routinely used in economic contexts to imply an inherently positive change. Strictly speaking, however, growth is nothing more than the quantitative increase in some measure of the economy—building permits, crude oil production, auto sales, and so forth. As such, growth signals only that something has become more or larger, not that the change is necessarily for the better. Within economics, it is proper (and better) to reserve for the word *growth* its strictly quantitative definition. As such, growth is not by itself good or bad—it is simply more of something.

An altogether different word, such as *development,* should be used within any qualitative context (referring to the achievement of some desired quality, e.g., the availability of health care security to all). Because of the common confusion of these terms, it is too often assumed that *growth* and *development* are synonymous. But growth does not guarantee development, nor does development necessarily require growth. Although development can serve as a valid social goal, it makes no sense to strive for growth for the sake of growth.

Yet that is precisely what seems to be happening, if one considers carefully the logic of economic measurement and goal setting in the United States (and most other market economies). The most widely reported measure of overall economic performance is an economic indicator known as gross domestic product, or GDP. Economists calculate GDP as the market value of newly produced goods, services, and construction within a country in one year. As such, it is purely a growth indicator of certain types of economic activity. It says very little about the well-being of people from a development sense—in psychological, social, or ecological terms. Every new mile of urban freeway built raises GDP, yet it also divides neighborhoods, generates additional air pollution, and encourages longer commutes at the expense of time spent with families. GDP rises, but is overall social welfare increased?

Nevertheless, GDP statistics have arguably become the singular indicator of overall government policy success for economists, reporters, voters, and politicians swayed by the illusion of equating growth and development. In recent years, painful government spending cuts have been justified in the name of facilitating growth, in spite of mounting evidence that many social development needs have now become immune to the remedy of increased production growth. In this light, it is not at all difficult to understand how the U.S. and world GDP can steadily rise while important development aspirations of societies everywhere (among them, safe streets and employment security) are not achieved.

Environmentally, GDP calculations fail even more miserably to reflect reality. They are blind to the negative effects of economic growth on the environment and especially to unsustainable rates of natural resource depletion. They make no distinction between desirable and undesirable environmental activities and tend to measure gain without also accounting for costs. For example, lumber produced sustainably on commercial tree farms is valued the same as comparable lumber clear-cut from one of the few remaining stands of old-growth forest. A bushel of Iowa corn is valued without taking account of topsoil lost to erosion in the cultivation process. The products of factories and farms that discharge chemical residues into public waterways are fully counted in GDP statistics, without being discounted for the losses to other life forms downstream. This problem knows no boundaries; DDT now shows up even in Antarctic penguins.[8]

In general, GDP growth rates accelerate whenever producers find ways to take free from nature whatever they can get and push cleanup costs into the future. Not only do GDP statistics fail to reflect environmental harm caused by such economic activity, they rise even further when money must be spent to repair the environment. GDP "grows" when hazardous waste is produced and then "grows" some more when money is spent on cleaning up chemical contamination, purifying water to make it drinkable, or treating the elevated incidence of cancer attributable to pollution.

It is thus apparent that GDP calculations not only mask the breakdown of the environment, they actually portray that breakdown as gain. Stated differently, much of what is routinely called *growth* is, in fact, merely the repair of past blunders. Recent attempts to recalculate GDP systematically—accounting for both the benefits *and* the costs of economic growth as they affect society and the environment—suggest that the GDP per American has hardly doubled since the 1950s, as official government statistics imply. Rather, the researchers' genuine progress indicator (or GPI) rose until about 1970 but has gradually declined by roughly 45 percent since then.[9] Herman Daly, a former World Bank economist and now a leader of the ecological economics movement, has concluded that "economic growth [may increase] environmental costs faster than it increases production benefits and thereby [make] us poorer rather than richer in an inclusive sense."[10]

The growth-equals-development myth will endure until more realistic accounting systems are devised and adopted for the routine generation of development progress reports. Such systems will track both benefits and costs of economic activity and account for changes in the quantity and quality of all forms of investment capital needed for a satisfactory future: financial capital, environmental capital, and social capital. Such information must be readily available if policymakers are to choose pathways leading toward the goal of sustainable development.

Myth 2: There Is No Limit to Growth

There are surely no boundaries to the human development of a creative nature. The mind can be sharpened, aesthetic beauty can be enjoyed without limit, and society can be enriched in countless creative ways. But a clear distinction must be made between these kinds of growth, which are without limit, and growth based upon the ever-expanding use of finite supplies of natural resources, which most certainly must have a limit.

In 1966, Kenneth E. Boulding wrote a short essay entitled "The Economics of the Coming Spaceship Earth," which has endured as perhaps the most eloquent statement of the realistic limits to growth. "We are now in the middle of a long process of transition in the nature of the image which man has of himself and his environment."[11] That image has, since early civilization, been of life on a virtually limitless plain without boundaries, of a frontier always available for a fresh start when social or environmental conditions became difficult elsewhere. Now, however, in a world of many billions of people, we are being forced to come to terms with the reality that the frontier is no more.

The "cowboy economy" of the past—"associated with reckless, exploitative, romantic, and violent behavior" toward nature—must be replaced with something that works in a closed sphere of limited size and resources. Suggested Boulding, we now must learn how to live in a "spaceman economy" in which "the earth has become a single spaceship, without unlimited reservoirs of anything, either for extraction or pollution."[12] Practically, this means that linear models of ever-expanding economic growth are no longer possible; future models of the economy will need to be circular and self-renewing.

More recently, economist Herman Daly and process theologian John B. Cobb Jr., in *For the Common Good*, posed the question: How big can the human economic subsystem be in relation to the natural ecosystem as a whole? They suggested an image of a growth-oriented economy quite literally "filling up" the earth's natural life-sustaining capacities. Contemporary economic models cannot capture this problem, they suggested, because they contain a common flaw: They simulate a human economy that contains the natural ecosystem, rather than the other way around. This error of perspective is of the magnitude of the difference between Ptolemy and Copernicus—is Earth or the sun the center of the universe?[13]

As a result of this flawed logic, the human growth-oriented economy claims for its own use ever-greater shares of nature's fixed supply of resources and dumps back into the earth's "environmental sinks" residues that exceed nature's absorptive capacities. The consequence is a planet that is exhibiting increasing difficulty in sustaining life for all species, humans included. There is dramatic evidence of declining fish stocks and fish fertility in bodies of water that have nurtured civilizations from ancient times—the Black Sea, Yellow Sea, Baltic Sea, Caspian Sea, and Bering Sea.[14] Throughout the world, unprecedented quantities of carbon are being released into the atmosphere as oil, coal, and wood are burned to meet energy demands. In sub-Saharan Africa, 80 percent of drylands and rangelands show significant signs of desertification (extreme loss of vegetative cover). As Daly and Cobb concluded, a fundamental problem of the growth-driven economic model in its demands on the biosphere is that the market has no built-in tendency to grow only to the optimal aggregate resource use (or even merely sustainable resource use).[15] As is widely recognized, the market does have internal incentives that seek *efficiency* of resource use, but there is no mechanism ensuring that efficiency occurs at the scale of the human economy that is best for all forms of life (including humans) in the long run.

How should the world's forests, meadows, plains, and seabeds be used? Already, it is calculated that 40 percent of terrestrial photosynthesis (from which the world derives its entire food supply) is appropriated by human beings in their various economic activities, at the expense of all nonhuman species of life.[16] Today, through the clear-cutting of rain forests and the draining of wetlands, the earth is experiencing a general rate of species extinction one thousand to ten thousand times greater than the background level of natural extinction that had existed for the past sixty-five million years of the Cenozoic Age.[17]

Some expansion of economic activity is undoubtedly necessary just to secure adequate food and shelter for the world's growing population, but as Arne Naess and George Sessions have proposed, "Humans have no right to reduce this richness and diversity [of nature] except to sat-isfy *vital* needs."[18] In this regard, American-style growth-oriented con-sumerism within a spaceship economy is viewed as both unethical and unsustainable.

A parallel to shifting more and more resources from other life forms to humans is found intergenerationally within human society. One of the key features of present-day economic growth is that it can all too easily come at the expense of future generations. Modern extractive technolo-gies make it possible for one generation to take as much as possible "free" from nature, leaving future generations much less of nature's bounty to work with. Can it really be said that a nation is growing if it achieves that growth by cutting down its centuries-old trees for quick sale or rapidly depleting its finite reserves of oil? "The notion of sustainability is about our obligation to the future," notes Nobel Prize-winning econ-omist Robert Solow.[19] Sustainability has become a matter of distribu-tional equity between the present and the future. Current generations must stop free-riding at the expense of the future.

The issue of intergenerational equity of natural resource use is rel-atively new, especially in the crucial realm of fossil fuel supplies. Before the dawn of the industrial revolution, economic growth originated pri-marily from use of renewable energy resources readily found on the sur-face of the earth—wood, animal power, water power, and wind power. Farmers were limited to utilizing the soil's natural fertility, which could be enriched with organic materials from other farm operations. Not sur-prisingly, the rate of economic growth was minuscule by today's stan-dards, but it was sustainable, since it originated from constantly renew-ing sources.

Contrary to popular thought, the great economic growth transfor-mation of the industrial revolution was due not so much to the use of ma-chinery as to a shift from the use of current to stored energy sources.[20] No longer was most energy obtained from surface mining of renewable resources. Rather, industrialists began mining subsurface, nonrenewable resources—fossil fuel energy that had been stored within the earth over millions of years. Thus, the real genius of the industrial revolution (if it can be called that) was to figure out how to use up quickly energy that had been stored from the sun over very long periods. Today, fossil fuel consumption per person is nine times higher in industrialized countries than in developing countries.[21] As we enter the twenty-first century, we continue to perpetuate a modern economic and technological system that burns in twenty-four hours an amount of energy that the planet took twenty-seven *years* to create.[22] This sort of economic development is without question unsustainable.

Of course, it can be argued that underground resources, in particular, are of no value if they are not put to productive use. By any objective standard, however, the rate of depletion of these resources—especially petroleum reserves—cannot be justified when considering the reduced set of energy options that will be available to future generations. The same concern applies to rates of deforestation and ground-water pumping in many parts of the world. In the United States alone, every day farmers and ranchers draw out from the ground twenty billion more gallons of water than are replaced by rainfall, losing priceless topsoil to salination in the process of irrigation. One important example of this reality is the enormous Ogalala Aquifer, which greens crops on much of the Great Plains and which will dry up within thirty to forty years at present rates of extraction.[23]

Lester Brown, founder of the Worldwatch Institute, portrays resource use patterns such as these as "a disguised form of deficit financing," which allows us to live above our means today at the expense of our children. Once again, the current system of GDP accounting plays an important role in this self-deception. The nation's official economic statisticians count only gains in output, without subtracting for the destruction of natural capital associated with those gains. Under such a system of economic accounting, the more that can be taken from nature for "free" and the more quickly it can be taken, the faster the measured growth rate of GDP. No business that wants to survive over the long run can afford to ignore in its financial statements the depletion of its productive assets, yet that is precisely what the global economy is collectively doing. Business ecologist Paul Hawken concludes, "Planet Earth is having a once-in-a-billion-years carbon blowout sale, all fossil fuels priced to move."[24]

In recent years, the problems of economic growth and environmental depletion have gained considerable visibility on the world's policy discussion stages. Most notable of these discussions was the 1992 United Nations Conference on Environment and Development in Rio de Janeiro (popularly known as the Earth Summit), attended by more than 30,000 people, including 118 heads of state and 9,000 journalists.[25] Despite the fact that environmental issues are now accepted internationally as perhaps the most pressing problem of our age, the official international institutions that guide our economic progress have not responded with binding, solutions-oriented policy changes.

In part, this is because many seem not yet convinced that environmental constraints are real. Often, concerns about environmental limits are simply dismissed by growth advocates with a faith best described as "technology to the rescue." The central tenet of this faith is that modern science has the capacity to find a solution to every problem as it develops. Evidence in support of this belief does indeed seem abundant. Global food supplies have managed to stay ahead of exploding population growth, thanks to the "green revolution" in high-input agriculture. Microcoating technologies have vastly reduced the need for conductive metals such as copper and gold in electrical applications. Oil shortages predicted with regularity since the dawn of the petroleum age have always failed to materialize as additional fuel efficiencies and new extractive technologies have been perfected.[26]

Inseparable allies of the "technological fix" school of thought are freely functioning markets. As every beginning student of economics is taught, shortages or supply difficulties of any sort tend to force product prices up. Higher prices both induce consumers to use less and stimulate sellers to find ways to produce more. Thus, the market solves the supply problem by rewarding both conservation and innovation in the face of scarcity.

Eventually, as emerging natural resource scarcities push prices higher, it will become economically profitable even to use previously impractical production technologies or product substitutes. In fact, the laws of supply and demand ensure that the world will *never* completely run out of oil or any other natural resource—long before the last barrel is pumped from the ground, the price will have risen to the point that no one will want to buy it. Thus, argue those who question the urgency of environmental limits, technological advances and market forces acting in tandem can readily handle any problems as they arise. It is prudent, they conclude, to pursue an agenda of nonstop economic growth even within a world of finite natural resources.

Such logic breaks down at several levels. First, faith in the techno-logical fix must itself be questioned. As Daly and Cobb note, "It is one thing to say that knowledge will grow (no one rejects that), but it is something else to presuppose that the content of new knowledge will abolish old limits faster than it discovers new ones."[27] The unrealized promises of virtually limitless, safe, nuclear energy are a prime example of a technology that, in fact, seems to create as many problems as it solves (see Brubaker's comments on nuclear pollution in chap. 2, for ex-ample).

At the center of many other technological solutions are synthetic chemical compounds released into the environment by the hundreds of millions of pounds annually. The multitude of solvents, fungicides, pes-ticides, and refrigerants made from the combination of chlorine and hy-drocarbons (the organochlorine family of compounds) persist in nature for decades, even for thousands of years. Yet they cannot be incorporated into the life cycle of any organism on earth. They are building up in the environment and steadily accumulating in our water, our food—and in the fatty tissues of our bodies. The multigenerational health effects of such chemically based technology cannot be determined for years to come, but it is already clear that organochlorines play a role in the ris-ing incidence of cancer, infertility, immune suppression, birth defects, and stillbirths.[28] It is not at all clear that the benefits of this technology outweigh its costs.

A second critique of the technology/market solution to environ-mental management focuses on the doors that, over time, *close* to future generations, not just the ones that are presumed to *open*. The open-door enthusiasts argue that our moral obligation to future generations is only, as Robert Solow expresses it, "to conduct ourselves so that we leave to the future the option or the capacity to be as well off as we are."[29] With such a goal, it becomes possible to argue that using up all the world's supply of aluminum or spotted owls isn't bad *as long as the technological capacity exists* for future generations to substitute other resources and maintain growth capacity. The implied understanding is that there are no limits that cannot be taken care of by substituting plentiful human and financial capital for scarce nature as it is used up.

Aside from the unknown future effects of technological fixes de-scribed above, such disregard for the sanctity of the created order is alarming for ethical reasons. By what right do human beings put to use their powers as the only species capable of eliminating another species or altering the natural environment? By what right does one generation, in the interests of short-term gain, deliberately live in ways that generate *irreversible* consequences to the future?

It may be true that technology can be used to "maintain capacity" in the future, but natural options are forever being closed by humans lacking sufficient wisdom or justification for taking such drastic steps. A case in point is the ongoing mass destruction of both plant and animal genetic diversity worldwide. At the present rate of extinction, 20 percent of the species on the planet may be lost within the next twenty to forty years, most of them in the tropical rain forests.[30]

In the growth-oriented marketplace, it is difficult to find a price on a resource deemed too high to pay for the sake of more rapid economic expansion. Natural resource use decisions—especially those with irreversible consequences—must *not* be ceded by society to private firms and individuals operating under the calculus of short-term financial gain or to a price system that cannot make reasoned judgments about relative values or purposes.

A final criticism of the technology / market solution to environmental management is one of distributional fairness. As is the case any time the market serves as the allocating mechanism of scarce resources, those with superior financial means are greatly advantaged in obtaining a share of the resources. People with fewer financial means inevitably lose out in the bidding to acquire use rights over the scarce resources. This reality of the market mechanism holds within rich and poor countries alike.

It is possible, in some circumstances, to justify a wealth-based allocation of human-made resources. But by what right can the resources of the *natural order*—provided by the Creator for the benefit of all—be claimed disproportionately by a few at the exclusion of others both now and in the future? One way of thinking in specific terms about how our actions asymmetrically appropriate the productive capacity of the earth and its resultant destruction is to use the analogy of an ecological footprint.[31]

In arid and sensitive environments, like the polar region of North America, footprints of men and machines can remain etched on the landscape for decades. For every good or service, a certain amount of productive land somewhere on the earth is appropriated as a source of raw materials and to assimilate a portion of the wastes we generate. Approximately one-quarter of the world's human population accounts for more than three-quarters of the total global consumption of energy and resources. The ecological footprints of North Americans are the largest in the world; the footprint of a Canadian averages 10.6 acres, and that of a resident of the United States is 12.6 acres.[32]

Each new child born in the United States, Canada, or Europe will deplete as many of the world's limited natural resources as will sixteen children in underdeveloped countries.[33] In contrast, the ecological footprint of persons in India is one-tenth the size of the average North American's. If everybody lived like North Americans, at least two additional Earths would be needed to produce the necessary resources, absorb the wastes, and maintain essential ecological processes.[34]

However, there is no consensus that wealthy countries should reduce their total load on the environment. As noted by Jocele Meyer, a participant at the Creation Summit, "There are many weapons used against creation that we as a society are not ready to renounce." Relying solely upon the market to determine access to natural resources raises serious distributional questions of fairness.

An alternative to market allocation is to make more natural resource use and access decisions through the democratic political process. Widespread public access to nature can be safeguarded at the same time that policy incentives can be deliberately tilted in favor of fuel efficiency, rational land use patterns, and resource sustainability. The political process itself may be imperfect. Even more problematic, however, would be to leave irreversible natural resource use decisions to a market mechanism efficient at producing growth but with no sense of the optimal scale of the economy relative to the ecosystem.

Myth 3: Growth Resolves Poverty

Some 1.2 billion people on Earth—one of every five—now meet former World Bank President Robert McNamara's definition of absolute poverty: "A condition of life so limited by malnutrition, illiteracy, disease, squalid surroundings, high infant mortality, and low life expectancy as to be beneath any reasonable definition of human decency."[35] This unfortunate reality is often used as a justification for giving overwhelming priority to economic growth policies—trading nature for growth, or spotted owls for more jobs, if necessary. It is argued that physical manifestations of poverty invariably require materially based solutions. Adequate housing, secure food storage facilities, and even basic medical treatment do, in fact, require increased resources at the disposal of the poor. But is the current global model of economic growth likely to achieve this goal?

There is little evidence that the fruits of decades of overall economic growth at either the global or the national scale have "trickled down" to the poor. Instead, wealth-poverty gaps seem to be growing, while the number of the world's poor continues to rise. Thirty years ago, the richest one-fifth of the world's citizens received thirty times more income than the poorest one-fifth—a 30 to 1 ratio. Now, in a much larger globalized economy with increased market integration and centralized power, the richest one-fifth receive sixty times more income than the poorest one-fifth—a 60 to 1 ratio.[36]

In the United States, the same phenomenon is occurring. In spite of the strong overall record of economic growth to the present, median household incomes, adjusted for inflation, have eroded steadily since their peak in 1973. Most Americans are working longer hours for less pay, and Americans now lead the industrialized world in the rate at which they moonlight on second jobs.[37] At the same time, the richest 20 percent of Americans is accruing fortune from growth without precedent. Between 1977 and 1989, the top 1 percent of households in the United States *by themselves* captured 60 percent of the nation's substantial increase in after-tax income.[38]

It is not surprising that the poor have failed to benefit from global economic growth. We have entered an era of jobless growth, in which technology and reorganization are eliminating good jobs faster than growth is creating them in the affluent countries. Skilled workers get pushed down the job ladder, causing those at the bottom to fall off entirely. In underdeveloped parts of the world, transnational industrial or agricultural corporations hire routine production workers—but fewer than might be expected, given their heavily mechanized production processes. The gains from trade are all but absent where they are needed most; the poorest 20 percent of the world's population is involved in barely 1 percent of world trade.[39]

The lack of "trickle-down" is not the only reason to question the efficacy of material growth as a strategy for poverty reduction. No rational solution to poverty can include adherence to an economic growth model preprogrammed to lead to disaster in the long run. Imagine the ecological consequences if the nearly two billion poor citizens of India and China would find a way to live a Western consumer's lifestyle—complete with disposable "convenience" products and two cars in every garage! Disaster looms precisely because the current economic model has no built-in limits—no stopping point short of a crisis generated by environmental or social collapse.

At the root of this phenomenon is the Western value of materialism, which comes packaged with prevailing economic policy assumptions. Specifically, Western materialism has no concept of "enough." More is always assumed to be better than less; even better, "you can have it all and you can have it now." Psychologist Erich Fromm senses the tragic consequences that this modern package of materialism holds for the world's poor: "In a culture in which the supreme goal is to have, . . . it would seem that the very essence of being is having. . . . [Thus,] if one *has nothing*, one *is nothing*."[40] Development ethicist Denis Goulet equates this mistaken modern Western value with "anti-development, which destroys nature, human cultures, and individuals and exacts undue sacrifices."[41]

In their book *Beyond Poverty and Affluence,* two Dutch economists, Bob Goudzwaard and Harry de Lange, call for a change in Western values:

> It involves developing a way of life that is content with "enough" and that demonstrates this contentment by a conscious acceptance of a level of income and consumption that does not escalate . . . It involves the realization that because of our collective drive for more and more, we directly damage our *own* well-being . . . The implementation of such a vision will create new possibilities for *neighborliness,* for demonstrating *care for our surroundings,* and for having more *time* available in our harried lives. Such a vision will help to liberate not only the poor but also the rich.[42]

What is the principal barrier to adopting a new economic paradigm of "enough" material development, which would greatly reduce the escalating pressures on the earth's environment? Why is economic growth so predictably the preferred antidote to poverty chosen by official policymakers of the world's ruling economic institutions? David Korten offers these answers in *When Corporations Rule the World:* "A belief in the possibility of unlimited growth is the very foundation of the ideological doctrine of corporate libertarianism, because to accept the reality of physical limits is to accept the need to limit greed and acquisition in favor of economic justice and sufficiency. Growth would have to give way to redistribution and reallocation of environmental resources as the focus of economic policy."[43]

In other words, as long as one espouses the possibility—realistic or not—of future economic growth, difficult policy choices can be avoided. It is only too easy to deal with thorny social problems and needed environmental expenditures by appealing to patience and the palliative of promised economic growth to come. Any pressure to modify existing institutional structures or the rules of the economic game itself can be fended off (for now).

But the environmental limits to growth confronting the human economy are threatening that time-tested strategy. Throughout human history, growth and expansion have always been the primary safety valves in managing the contentious politics of rich and poor. An environmentally "full earth" will require, for the first time, that the root systemic causes of poverty be addressed much more directly. The alternative of global environmental collapse would not be pleasant for anyone, rich or poor.

In many parts of the world, desperately poor people who lack either an available environmental frontier or the financial means to protect their existing natural resource base are resorting to acts of deforestation, overgrazing, overfishing, and industrial spoilage. In part, they are doing this because their governments have been forced by the International Monetary Fund and World Bank to expand production at any cost to facilitate payments on the impossible foreign debt burdens accumulated under the current international financial regime.

To these strains on the global ecosystem are added the even more significant environmental damage of the world's affluent and their insatiable throw-away consumerism and run-away fossil fuel demands. The United Nations Conference on Environment and Development estimated in 1992 that 75 percent of all global environmental damage is caused by the 25 percent of the world's population who live in the developed countries of the North.[44] Northern Hemisphere "advanced" countries have been pressing southern countries to reduce destruction of valued forests and biological resources, while the "less advanced" in turn have been striving for the benefits of unfettered industrialization and economic expansion.

However, comparisons of per capita consumption do not adequately reflect our relative demands on Earth's productive capacity. Per capita consumption does not consider, for example, the physical area being appropriated by daily activities and consumption patterns.[45] The natural environment is clearly approaching its human carrying capacity. This is occurring as much for reasons of excessive material consumption by the world's affluent as for reasons of growing populations among the world's poor. Each year, ninety million people are added to the world's population, mostly in the Third World.

Every major indicator shows deteriorations in the world's natural systems.[46] If problems of too much consumption and too much poverty are to be addressed—as they must be to avoid even greater environmental damage and political instability in the future—policies other than simply "more economic growth" must be pursued. As the massive attendance at the 1992 Rio "Earth Summit" demonstrated, such conversation is indeed beginning. The need to sustain our one common heritage, the environment, may yet prove to be the power that forges new and positive links among all peoples of the world, none of whom can secure their environmental futures alone.

THE GLOBAL ECONOMY AND BARRIERS TO ENVIRONMENTAL HEALING

As urgent a need as worldwide environmental cooperation has become, the ongoing process of economic globalization is adding new environmental dangers of its own. The new global economy is signified by the disappearance of national markets and the creation of a single international marketplace for rapidly expanding trade and foreign investment. It is characterized by a consolidation of economic power into fewer and fewer hands, often distantly located from those who are affected by their decisions. The world's 500 largest transnational corporations effectively control 70 percent of all world trade. At the individual level, the world's 358 billionaires who reap their fortunes from these enormous markets now have a combined net worth equal to that of the poorest 2.5 billion citizens of the world.[47]

On the surface, this concentration of power into readily identifiable hands might seem like an encouraging development for greater international environmental cooperation. In reality, it all too often serves as a mechanism for perpetuating the short-term abilities of the world's affluent citizens—the primary beneficiaries of the global economy—to continue a lifestyle above their environmental means. The global economy concentrates power in two ways: through free trade and through unrestricted cross-border investment. The successful push in recent years by the world's leading economic powers within the World Trade Organization toward the eventual worldwide elimination of tariffs, quotas, "red tape," and other export/import restrictions has made it easier to obtain coveted natural resources from anywhere in the world. Natural resources can be imported more readily in any form—unprocessed or as intermediate or finished goods. The result in the global marketplace is that those with superior purchasing power are able to outbid everyone else for nature's bounty.

The 20 percent of the world's population who live in the richest countries can effectively use global markets to get whatever natural resources are available, since they control nearly 80 percent of the world's total income and purchasing power. Accordingly, the best original-growth lumber from anywhere in Southeast Asia (and the American West Coast, for that matter) inevitably ends up in the hands of Japanese businesses. Oil is always available in the United States, even for recreational uses, when citizens of oil-exporting countries themselves must sometimes go without it.

Many countries would not be able to support their current resource consumption levels were it not for trade. To meet its food and timber demands alone, the Netherlands appropriates the production from twenty-four million hectares of land worldwide—ten times its own area of cropland, pasture, and forest.[48] Superior purchasing power makes it possible for this small country to draw upon the ecological carrying capacity of other nations. Some of its coffee supply, for example, comes from the fertile Kenyan highlands. Farmers in that land-short region, like elsewhere in the global economy, now work for the highest bidder worldwide. They have stopped producing food for Kenya's own rapidly growing population because Dutch morning coffee drinkers are able to pay more for the use of Kenyan land than can Kenyans. Coffee trucks, carrying nature's bounty, pass by hungry children on their way to the port of Mombasa, where the coffee is loaded onto ocean freighters for shipment to Europe. By the same market mechanism and in the face of 786 million undernourished people worldwide,[49] car owners can appropriate fertile land by fueling their cars with alcohol made from food crops. The affluent will be the last to comprehend the reality of any impending natural resource shortage.

As Andrew Schmookler describes this phenomenon, "The market— no less than the political systems of kingdoms—is a power system and as such can grant disproportionate power to those people . . . [who] are most able to orient themselves according to the system's lines of force."[50] In a free-market global economy, each individual is in competition for access to the world's limited environmental space, and the person with the most money invariably wins.

The lifting of foreign investment barriers under recently liberalized trade rules is the second mechanism by which the world's economically privileged are able to sustain a lifestyle above their own environmental means. Economic globalization has greatly expanded the opportunities for the rich to pass their environmental burdens to the poor by exporting both polluting factories and wastes. Multinational corporations can relocate environmentally damaging production facilities away from their home bases—often to poor countries desperate for jobs at any cost. The relentless pressure of price competition in the global marketplace is the force ensuring that any cost-reduction potential will be exploited. In the worst cases, environmental rape is teamed with sweatshop labor to produce goods for export to affluent countries at bargain prices. Observes Paul Hawken, "In the upside-down and inverted logic of the present economic system, we cannot imagine there is a point where something is too cheap."[51]

Japanese-financed smelters now provide much of that nation's copper supplies from the Philippines on land expropriated by the Philippine government from local residents at give-away prices. Through this arrangement, Japanese citizens get a steady supply of competitively priced copper at no environmental costs to themselves. The Filipinos do gain some employment but must cope with the smelter's gas and wastewater emissions containing high concentrations of boron, arsenic, heavy metals, and sulfur compounds, which have contaminated local water supplies, reduced fishing and rice yields, damaged the forests, and increased the occurrence of upper-respiratory diseases among local residents.[52] David Korten assesses such global economic arrangements in harsh terms:

> Although an open trading system is sometimes advocated as necessary to make up for the environmental deficits of those who have too little, it more often works in exactly the opposite way—increasing the environmental deficits of those who have too little to provide a surplus for those who already have too much. Furthermore, an open trading system makes it easier for the rich to keep the consequences of this transfer out of their sight. The farther out of sight those consequences are, the easier it is for those who hold power to ignore or rationalize them.[53]

Some environmental problems cannot so easily be shifted elsewhere by trade or investment activities. Worldwide atmospheric changes such as global warming and ozone depletion are prime examples. In the case of global warming—predicted to occur as a consequence of high levels of greenhouse gas emissions into the earth's atmosphere—hotter surface temperatures, altered rainfall patterns, melting polar ice caps, and rising sea levels will affect the whole world. Yet it will be the developing countries, whose economic activities contribute less than 30 percent of all greenhouse gases, that will suffer the most serious consequences of global warming.[54] The affluent nations, which reap most of the economic rewards of global warming activities, are able to pay to *protect* themselves from the negative effects of their own pollution. Coastal cities, such as New York and London, will use expensive barriers and Dutch-style ocean walls, if necessary, to hold back rising sea tides in the middle of the next century. The same financial ability does not exist for impoverished coastal countries such as Bangladesh, where most residents already suffer from the effects of periodic flooding.

The environmental policy calculus of affluent nations and global corporations hesitant to adopt strict greenhouse gas reduction targets seems to be that it is cheaper to adapt to climate change than it is to stop causing it. Such a "rights by income" approach virtually assures that the world's poor will suffer disproportionate consequences of global warming. Those who transact less in the global economy are victimized more, because they must bear the burdens of negative market externalities without consuming the benefits.

AN ECONOMIC ETHIC THAT CARES
FOR THE ENVIRONMENT

Maurice Strong, the outspoken Canadian secretary general of the 1992 United Nations "Earth Summit," observes that "we have lost our innocence. We know what we are doing today to the environment that God has bequeathed to us as our endowment on this earth. We know what we are doing to future generations. We know what we are doing to each other. And our knowledge poses the ultimate moral and ethical challenge to the very foundations of our civilization."[55] Becoming honest with ourselves on these points is the first step toward the creation of a new economic ethic that motivates environmental stewardship. This ethic must bring an end to the destructive logic of the paradigm of unsustainable economic growth that dominated the twentieth century. To succeed, this ethic must change fundamental relationships of economic life at the individual level and, in so doing, heal the economic wounds inflicted upon both nature and society.

There are many opportunities for communities and individuals to adjust their ecological footprints. Footprints can be reduced through lifestyle choices and careful land use planning. Riding a bicycle creates a footprint that is about one-tenth as large as driving your own vehicle. By eating less meat and processed, prepackaged foods and by buying longer-lasting products, one's ecological footprint can be reduced threefold in the food and consumer categories.[56]

Resonance with the sentiments expressed by Maurice Strong—that something is seriously wrong with the economic path we are currently taking—has led many to offer perspectives on what a new public economic ethic must entail.[57] Central to all such analyses is a vision of an economy that must be more just, participatory, and sustainable. The new economy must distinguish needs from wants and require simple living in which consumption is limited. It must thrive on a feeling of sufficiency rather than unlimited desires. It must view nature through the lens of stewardship rather than ownership.

But can such a radical transformation of values be accomplished? Russian religious and social philosopher Alexander Solzhenitsyn suggests the following individual motivation for change: "We start . . . from what seems to us beyond doubt: that true repentance and self-limitation will shortly reappear in the personal and the social sphere, that a hollow place in modern man is ready to receive them . . . After the Western ideal of unlimited freedom, after the Marxist concept of freedom as acceptance of the yoke of necessity—here is the true Christian definition of freedom. Freedom is *self-restriction!*" Solzhenitsyn's recognition of the liberating qualities of self-restriction diverts energy from outward to inward development efforts. He terms the adoption of the principle of inward development a "great turning point in the history of mankind," after which the individual "will not flog himself to death in his greed for bigger and bigger earnings, but will spend what he has economically, rationally and calmly."[58]

Yet there is strong evidence that materially oriented outward development will not be readily abandoned by individuals caught up in a global market system that so effectively nurtures the desire for it. To avoid collective environmental catastrophe, we must find a corrective to the myopic global economy. Its most harmful results can be reduced by increasing the frequency with which economic decision makers—consumers as well as producers—rub shoulders with those who must cope with the distributional and environmental consequences of their actions.

Daly and Cobb are among those who argue that the global economy's blind spots can be addressed by individual and public efforts to restore a sense of economic "community" at the local level.[59] Wherever possible throughout the economy, largeness and distant control must be made to give way to smallness and local control. This, in turn, will create manageable zones of mutual accountability and responsibility for self, others, and natural surroundings. Seldom is poverty found where these conditions are met, anywhere in the world.

Within communities, a sense of place and belonging frees individuals to let go of narrowly defined self-interest in favor of achieving society's broader goals. Life takes on a new, deeper set of meanings, fostering an ethic of inward development and voluntary material self-restraint. A spirit of mutual aid and heightened concern for environmental stewardship is a natural outgrowth of life in community, where people expect future generations to live long-term with the consequences of current decisions.

For our environmental future to be secured, the economic pendulum must swing back from the impersonal, individualistic global economy toward an era of renewed cooperation within strengthened local communities. Insofar as humans desire environmental justice, we can still make a difference.

CHAPTER 2

Science, Technology, and Creation

Kenton K. Brubaker

Some technologies ought not to be used under any circumstances, because what they do threatens either the sacred quality of life or the survivability of life.
—Jeremy Rifkin, *Declaration of a Heretic*

I remember distinctly the day I sat listening to a lecture by Jeremy Rifkin, when we considered the environmental and health effects of nuclear technology. Rifkin enumerated various aspects of the nuclear industry: nuclear weapons testing, waste disposal problems, atmospheric and ground-water pollution, and health hazards. The Three-Mile Island and Chernobyl disasters were in our consciousness. Rifkin asked us to raise a hand if we agreed that we could do without this amazing technology. It was like a religious revival meeting; I raised my hand.

The impact of scientific discovery and resultant technology on the natural world has been staggering. A brief, historical survey of technology and its effects on the planet might include toolmaking and the use of fire (pyrotechnology), settled agriculture replacing hunting and gathering, the steam engine and the industrial revolution, the Green Revolution and the introduction of high-yielding hybrid crops, nuclear energy, genetic engineering, microelectronics and communications, and nanotechnology. In this chapter I deal briefly with the last four, the more modern technologies in this list: nuclear energy, genetic engineering, communications, and nanotechnology.

Before we look at these modern technologies, consider the three surges of population growth and their correlation with technological revolutions in the past. Deevey is very helpful in describing these remarkable effects, for he points out that, with the advent of *toolmaking* about 1 million years ago, the human population increased about 33 times, from about 150,000 to 5 million.[1] Population then remained on a plateau until the development of *settled agriculture* about 10,000 years ago, resulting in a 100-fold increase in population to 500 million, which then again leveled off. The advent of the *industrial revolution* in the eighteenth century led to the current world population of more than 6 billion, a 10-fold increase. This most recent human population surge, however, occurred in just 300 years, while the other two increases happened over millennia.

Of these three technological revolutions, agriculture remains the most remarkable. It has been estimated that humans have now diverted half of the photosynthetic capacity of the planet to the support of *Homo sapiens.* Many of us believe that we cannot sustain the intensity of agriculture needed to support this growing human population without irreversible damage to soil and water systems.[2]

NUCLEAR ENGINEERING, VIOLENCE, AND ENVIRONMENTAL POLLUTION

The splitting of the atom and all the related discoveries in *nuclear engineering* remain one of the most creation-threatening technologies of our century. On December 2, 1942, a small group of experimental physicists carried us out of the age of fire into the age of nuclear power. E. F. Schumacher gives these dire warnings about nuclear energy: "Radioactive pollution is an evil of an incomparably greater 'dimension' than anything mankind has known before. The continuation of scientific advance in the direction of ever-increasing violence, culminating in nuclear fission and moving on to nuclear fusion, is a prospect of terror threatening the abolition of man."[3]

Schumacher warns that plutonium is one of the most deadly and dangerous products of the nuclear industry. An amount the size of a grapefruit would be enough to destroy the world. Yet we continue producing and storing this awesome substance, hoping that it will not fall into the hands of terrorists or unfriendly nations. Schumacher pleads for a radically different direction of scientific research, "towards non-violence rather than violence; towards a harmonious cooperation with nature rather than a warfare against nature; towards the noiseless, low-energy, elegant, and economical solutions normally applied in nature rather than the noisy, high-energy, brutal, wasteful, and clumsy solutions of our present-day sciences."[4]

The Hanford, Washington, plutonium enrichment site has "metastasized into the most polluted place in the Western world."[5] The 560-square-mile federal reservation is home to two-thirds of the country's high-level waste, buried or stored in tanks, a source of contamination to ground water and the Columbia River. The most polluted sites of the reserve can never again be safe for human life. Hanford manufactured 53 tons of plutonium before it ceased operation in 1987. For every kilogram of plutonium product, the plant generated 55,000 gallons of low- to mid-level radioactive waste, dumped into dirt trenches, and 340 gallons of high-level waste, pumped into underground steel tanks.

Ironically, Hanford is "a fine place to see an eagle hunt, a salmon spawn or a deer graze. But best not drink the groundwater for a quarter-million years." Abundant wildlife is also well known at the other highly contaminated U.S. nuclear facility, the 310-square-mile Savannah River Site in South Carolina. Here bomb makers spilled radioactive contaminants plutonium-237, strontium-90, and cesium-137 "all over the place, irradiating not only water and soil, but fish and fowl. Despite the fact that the sunfish have strontium-90 in their bones and the wading ducks emit gamma rays, the Savannah River Site boasts some of the healthiest and varied populations of birds, reptiles, amphibians and fish in the Southeast."[6]

So, even though we strongly lament the effects of radioisotope pollution on the human species, wildlife protected from people does seem to thrive, even in polluted environments. On the face of it, these developments raise rather interesting and possibly contradictory questions. Ecologists seem to attribute this paradox not so much to the effects of radiation as to the absence of people and consequent development. This emphasizes the position of many environmentalists, who argue that nature probably suffers more from human activity, such as habitat destruction, than from some of the pollutants we produce. In any case, we need more data to be sure of the long-term effects of environmental pollution before we dismiss radiation and related pollution as benign.

GENETIC ENGINEERING AND THE DISAPPEARANCE OF MATING BOUNDARIES

Genetic engineering is based on the discovery of molecular knives called *restriction endonucleases*, which H. O. Smith and others, in the late seventies, used to cut DNA at specific sites. These segments of the genetic code could then be isolated, cloned, and moved about within an organism or between organisms, even organisms of different species. This powerful biotechnology enabled technicians to introduce genes from almost any source into any other living organism, thus avoiding the normal mating barriers of nature. The revolutionary impact of this new knowledge and creative power brings into focus the questions, Why is there such a thing as species? What is the role of mating barriers in nature, and if we violate them, what will be the consequences?

Nature has been scrambling to invent new genetic material for millennia, but always within certain barriers, which we have come to recognize as species. By definition, organisms are considered to be different species if they do not successfully interbreed under natural conditions. Occasionally, new species arise, but these again are reproductively isolated, by definition. Even though we can now manufacture genes to our own specifications, those that have been invented and tested in nature are usually superior. For this reason we wish to preserve wild plants and animals; they are ancient and tested reservoirs of genetic wisdom. Our new ability to move, to invent, and to "correct" genetic information represents a major invasion of the natural system. The whole concept of species is somehow threatened. If the natural world respects boundaries within which the evolutionary process operates, what will be the consequences of a world without boundaries? Do we have the wisdom needed to manipulate genes at will? Robert Lucky observes that "science urges us ever forward, but science alone is not enough to get us there."[7] Is "forward" to be a natural world of gene pools without boundaries? In a sense, such a world has already arrived, at least potentially.

COMMUNICATIONS AND KNOWLEDGE

There is considerable concern that the artificial pace of modern communications, especially television, has produced a kind of alienation from nature. Fred Krueger points out that "nature moves really slow," while a medium such as television "synchronizes our internal processes with the artificial world of computers, freeways, air travel, and everything else that is fast-paced and accelerated. It attunes the human brain and nervous system to the lightning speeds of electronic systems." Krueger feels that the medium itself is as much the problem as inappropriate program content. He complains that television has the following effects, complaints that I share.[8]

1. Lessening of imagination, ability to play, and personal and social interaction
2. Predisposition to violence and hyperactivity
3. Tendency to leap over essential elements of analysis and judgment because of the accelerated pace and time constraints
4. Promotion of consumer mentality, instant gratification, and hedonistic culture
5. Decreased ability to focus and maintain attention
6. Trivialization of the natural world by reducing its wonders to sex and violence, thus distorting nature's rhythms, scope, and diversity

Communication by computer networking shares some of the characteristics of television, but in many ways is a completely new phenomenon. Nicholas Negroponte describes brilliantly what future communications will be like in his conventional book format, *Being Digital:* "The cost of the electronics in a modern car now exceeds the cost of its steel. It already has more than fifty microprocessors in it. That does not mean they have all been used very intelligently. You can be made to feel very foolish when you rent a fancy European car and realize when you are at the front of a long line for gas that you do not know how to electronically unlock the gas tank."[9]

One of the incredibly fascinating aspects of "being digital" is Negroponte's prediction of "interface agents" compiling information. Imagine a future in which your interface agent can read every newswire and newspaper and catch every TV and radio broadcast on the planet and then construct a personalized summary. This kind of newspaper is printed in an edition of one.

The VCR of the future will greet his arrival home by saying, "Nicholas, I looked at five thousand hours of television while you were out and recorded six segments for you, which total forty minutes. Your high school classmate was on the *Today* show, there was a documentary on the Dodecanese Islands, etc." (179). The VCR will scan the programs by looking at the headers.

Farther into the future we will be hybridized with the technology, even more so than those joggers I now see wired for sound. They will never know if Rachel Carson's "silent spring" has arrived, for they are listening to a different drummer: "In the further future, computer displays may be sold by the gallon and painted on, CD-ROMs may be edible, and parallel processors may be applied like suntan lotion. Alternately, we might be living in our computers" (211).

In an epilogue, Negroponte does acknowledge some of the downside of "being digital."

> The radical transformation of the nature of our job markets, as we work less with atoms and more with bits, will happen at just about the same time as the 2 billion-strong labor force of India and China starts to come on-line (literally).
>
> As we move more toward such a digital world, an entire sector of the population will be or feel disenfranchised. When a fifty-year-old steelworker loses his job, unlike his twenty-five-year-old son, he may have no digital resilience at all.
>
> Bits are not edible; in that sense they cannot stop hunger. Computers are not moral; they cannot resolve complex issues like the rights to life and to death.[10]

Does the increased speed with which we now communicate really produce a better world? Is there any evidence that more intelligence is being expressed? I doubt it. We still lack the wisdom and moral character needed to control the power this increased speed of communication gives us. Those who are outside the net or confused by it are at a greater disadvantage than ever before. The divergence between the haves and the have-nots continues to expand.

Negroponte is optimistic about our new abilities to compute and communicate in the digital era. He describes "four very powerful qualities that will result in its ultimate triumph: decentralizing, globalizing, harmonizing, and empowering."[11] I fear that this optimism is misplaced. The centralizing power of communications technology is best shown in our modern banking system. The small, independent bank is gone, replaced by impersonal megabanks whose operations leave the average person out in the cold. The global, harmonizing effect works equally well for terrorists and militia movements. Invasion of privacy and proliferation of scams, as well as widespread moral pollution, are all part of the downside of our current communication networks. Distribution of power and information are in no sense democratic; the poor and illiterate continue to be outside the loop.

Another caution that tempers my optimism about communications technology is the rapid rate at which hardware becomes outdated and discarded. Add to this the prodigious amount of paper wasted by malfunctioning printers and the need to redo reports because of poor editing capabilities, and we have some real problems to address.

NANOTECHNOLOGY AND IMMORTALITY

Nanotechnology involves the miniconstruction, atom by atom, of biological or physical structures. In *Scientific American*, Marvin Minsky does a wonderful job of describing a nanotechnology-designed world without limits.

Will robots inherit the earth? Yes, but they will be our children . . . Eventually, using nanotechnology, we will entirely replace our brains. Once delivered from the limitations of biology, we will decide the length of our lives—with the option of immortality—and choose among other, unimagined capabilities as well.

Suppose we wanted to copy a machine, such as a brain, that contained a trillion components. Today we could not do such a thing (even with the necessary knowledge) if we had to build each component separately. But if we had a million construction machines that could each build 1,000 parts per second, our task would take mere minutes. In the decades to come, new fabrication machines will make this possible. Most present-day manufacturing is based on shaping bulk materials. In contrast, nanotechnologists aim to build materials and machinery by placing each atom and molecule precisely where they want it . . . Our nanotechnologies should enable us to construct replacement bodies and brains that will not be constrained to work at the crawling pace of "real time." The events in our computer chips already happen millions of times faster than those in brain cells. Hence we could design our "mind-children" to think a million times faster than we do. To such a being, half a minute might seem as long as one of our years and each hour as long as an entire human lifetime.[12]

Nanotechnology, as envisioned by Minsky, seems to be the final rejection of mortality, time, and nature. Being "constrained to work at the crawling pace of 'real time'" does not seem to be too bad an idea to me. Such constraint helps give room for contemplation, reflection, and appreciation of the "real nature" of our world. Since the natural world crawls along, obeying all the intricate cycles of matter and energy flows, our own bodies and brains might want to remain in synchronization with these processes.

Minsky's concept of the brain as possibly made up of a trillion components is probably indicative of his limited concept of natural complexities. Surely, if we want to construct "atom by atom," the number of components will be trillions of trillions. To instruct machines to construct a brain or any other complex organ or tissue will be quite impossible because of the dynamic nature of these structures. From biology we know that such "machines" are self-constructed, using internal directions (DNA) responding to thousands of internal and external signals.

A final comment on Minsky's dreams of immortality. His humor comes through in the following observation: "I find it rather worrisome that so many people are resigned to die. Might not such people, who feel that they do not have much to lose, be dangerous?" No, we aren't dangerous. Being resigned to mortality is a sign of cultural maturity. People who cannot face their personal death are more to be feared. Our frenzied drive to fix every biological shortcoming has been a great boon to the medical and pharmaceutical industries, but there are limits to our obsession for immortality. A natural world without limits simply does not exist. There never will be enough kidneys and hearts to replace every one that wears out, even if we construct them using nanotechnology. There is no such thing as perpetual motion. I could use such an agent to screen my e-mail!

TECHNOLOGICAL OPTIMISM

Robert A. Wauzzinski deals brilliantly with the philosophical optimism of *technicism*, the idolatry of technology. He observes that technological optimists, such as Julian Simon and R. Buckminster Fuller, believe that a technical solution can solve any technically related problem:

> If cars are polluting, then catalytic converters can clean up the mess. To technology is added more technology. Accordingly faith [in technology] is demonstrated and augmented. We *must* place on-line—willy nilly—the latest and greatest technology because our natures and salvation are believed to be dependent upon it. The consequences, if negative, be damned; we can technically fix it later. Such is the spurious reasoning behind the advent of "peaceful" nuclear energy in this country.[13]

The danger of a "technologically saturated society" is that we attempt to locate technical solutions for nontechnological problems. Wauzzinski is correct when he notes that "future tragedies like those of Chernobyl and Challenger can be stopped when *humility* reduces the scope of technology, not by *adding* new systems."[14] His call for treatment of technical problems in concert with concerns about wholeness, justice, harmony, and respect for nature, all done with humility, demands careful consideration. Technological optimism, especially when it becomes technism, must be tempered with holistic wisdom and humility.

CORPORATIONS AND CONSUMERISM

Technology itself is not the source of the problem we are addressing; the problem is its misuse. One very plausible theory as to why we are engaged in such major misuse of technology is the role of advertising and corporations in fostering overconsumption. According to Frank Cougar, "The consumer culture only emerged as a consequence of deliberate efforts by retail corporations to reshape American values and to cultivate an artificial demand for products."[15]

Television accelerated the process of creating desire as the idea of aggressive marketing took hold. Bold colors, glamour, and exciting graphics made advertising more interesting than even the Super Bowl. Today, shopping on the Web promises instant access to every product needed to achieve a paradise of luxury and contentment. Newspapers, fliers, magazines, billboards, television, and now home-computer shopping all compete vigorously for the privilege of creating desire.

Corporations not only finance the media; they also have an increasing effect on the political process. Due to the enormous cost of getting elected, politicians "vote for continued expropriation of natural resources when the majority of their constituents want these resources preserved. Tellingly, many legislators are choosing to go with the money rather than their constituents."[16] Why is the corporation unable or unwilling to limit its exploitation of technology? Cougar discusses the following points:

1. Even ethical corporate managers are imprisoned by the corporate form. The corporation is an artificial being, not a real person, it doesn't have "feelings" or a heart, and so it has serious trouble discerning right from wrong.

2. Corporations have no size limits. They just keep growing far beyond what natural limits might allow.

3. Corporations are global, putting their decision making beyond national control, which means essentially no control.

4. Corporations exert overwhelming influence on government through massive campaign contributions, threatening the basic premise of government "by the people."

5. Corporations are very efficient at externalizing social and environmental costs so that only operating costs are passed on to the public.

6. Corporations are pouring money into shaping social ideas and values to counter what people think on their own. For example, they pose as "green" while privately supporting antiecological legislation.[17]

Technology out of control is not due primarily to the nature of technology itself; it is due more to the way in which society adopts and values the innovation. Institutions, such as the corporation, that do not have the wisdom or incentive to evaluate the effects of rampant technology cannot be trusted to regulate it successfully. It seems to be up to thoughtful, caring persons to sound the alarm when technology threatens the very civilization it pretends to produce.

CONCLUSION

Science and technology have given our planet an amazing capacity to support human life. Robert Kates observes that there has been a pronounced surge in human population with each major advance in human technology: toolmaking, agriculture, and industry. In asking whether life as we know it can be sustained on Earth, Kates says "the answer is almost assuredly 'no'."[18] His three areas of concern are pollutants (acid rain, heavy metals, and other chemicals), global atmospheric dangers (nuclear fallout, ozone depletion, climatic warming from greenhouse gases), and the massive assault on the biota (deforestation, desertification, and species extinction). Science and technology have allowed surges in human population, which now includes over a billion desperately poor people among the more than 6 billion total. It has become apparent to many that more technology will not remedy this situation. Kates suggests three crucial ideas that I affirm:

1. Cohabitation with the natural world is necessary. There needs to be ever-increasing concern for the environment.
2. There are limits to human activity.
3. The benefits of human activity need to be more widely shared.[19]

Cohabitation with the natural world is absolutely necessary. Our life comes from the natural processes of photosynthesis and materials recycling, powered by the sun. Any technology that destroys any of these natural systems cannot be employed. As world human population grows, these constraints become even more important. Interference with natural pools of genetic potential may prove to be disastrous. The free movement of genes between species could threaten the entire biosphere with disintegration.

There are limits to human activity. The persistent striving for immortality serves no useful purpose. Acceptance of our mortal place in the natural scheme of things is a mature and wholesome attitude. Animals and plants have their rightful place in the biosphere and should not be sacrificed casually for the deification of the human species. Acceptance of the natural pace of biological activity, including the limits of the human brain, leaves room for contemplation and the emergence of wisdom.

Given today's wide discrepancy between the rich and the poor, the privileged and the underprivileged, every effort should be made to narrow that gulf rather than contribute to greater disparity. Access to technology is power, and those who wield the power must show considerable constraint, especially toward those who will suffer the most as that power is manipulated.

Consumers must learn to resist the temptation to think that life consists in the abundance of things possessed (Luke 12:15). Corporate managers need to show restraint in the proliferation of society- and environment-damaging technology. Legislation should be strengthened to limit the power of corporate control over the political process. Educators need to explore with their students and colleagues the power and influence of technological optimism and technism. To recapture a sense of the sacred quality of life, we should contemplate again the complexity and diversity surrounding us in the natural world. We must value this complexity and diversity highly, more highly than speed, efficiency, and personal power. We need also to value social relations between people and aesthetic relations to beauty, especially the beauty of natural and human creations. Technology can be beautiful, but not if it destroys the sense of the sacred.

CHAPTER 3

Population Density
and a Sustainable Environment

Carl Keener and Calvin Redekop

A finite world can only support a finite population.
—Garret Hardin, "The Tragedy of the Commons"

When God "blessed Noah and his sons" and told them to be "fruitful and multiply and replenish the earth" (Gen. 9:1), He did not tell them how densely the earth should be replenished.[1] Ironically, the fate of our habitat—mother earth—is in large measure being determined by the numbers of humans (and animals) who inhabit it. During geohuman time (when humans have lived on and interacted with the planet), the numbers of people on the face of the earth have largely affected its fate.

POPULATION GROWTH AND THE ENVIRONMENT

Population density is one of the most obvious forces affecting nature, yet one of the least discussed. The growth of human population density throughout history has intensified its effects. In recent years the human population has begun to grow exponentially. "From the beginning of humanity's appearance on earth to 1945, it took more than ten thousand generations to reach a world population of 2 billion people. Now, in the course of one human lifetime—the world population will increase from 2 to more than 9 billion."[2]

However, population growth and density have never played a great role in determining the practices, beliefs, and politics of human society. The belief in the individual or personal right to human reproduction has apparently been one of the most emotionally defended of all human rights—hence, the lack of discourse regarding density of population and its effects on the environment.[3] But increased awareness of population growth is beginning to stimulate discussion of its implications.

The first famed expression of concern over uncontrolled population growth was articulated by Thomas Malthus in 1798.[4] A storm of controversy broke out in the religious, scientific, and philosophical communities and then for a time was rejected or ignored.[5] In the twentieth century individuals and groups began to reconsider the basic substance of Malthus's claims. Designated *neo-Malthusians,* they have warned that a crisis in uncurbed human population growth is emerging. One line of concern has been the carrying capacity of the earth: only a finite number of living things can exist sustainably on our globe.[6] One recent survey and analysis of the research suggests the range of projected population growth: "a medium projection of ten to eleven billion and a low-high range between eight billion and fifteen billion by the end of the next century."[7] Sixteen billion now seems to be the generally accepted maximum for human life on the planet.[8]

The prevailing discussion of how many persons constitute overpopulation almost totally ignores two critical and major factors: (1) the destruction of natural habitat and the environment by the incursion of human populations on the wilderness and (2) the quality of human life. When the welfare of the entire ecosystem is taken as the basic starting point, the world is already vastly overpopulated. If sustainability alone is the criterion, Anne and Paul Ehrlich maintain that "the entire planet and virtually every nation is already vastly overpopulated."[9] If we take the rights of the plant and animal kingdom into consideration in our environmental analysis, it seems apparent that disaster is already upon us.[10]

On the other side of the discussion, *optimistic cornucopians* have confidently maintained that science and technology and the personal desire for self-improvement will finally solve all problems associated with increasing population density, without reference to wilderness rights or quality-of-life issues. Donald Bogue triumphantly predicted that "the world population crisis is a phenomenon of the 20th century, and will be largely if not entirely a matter of history when humanity moves into the 21st century. It is probable that by the year 2000 each of the major world regions will have a population growth rate that is either zero or is easily within the capacity of its expanding economy to support."[11] Ergo, the concern with overpopulation is folly. The optimists would say that there is no evidence to be drawn from any of the demographic literature with implications for the environment such as pollution or destruction of the natural habitat, as though population density is unrelated to the physical world (a curious contradiction of the basic tenet of the concept of ecology, namely, that everything is affected by every other thing).

POPULATION DENSITY AND THE ECOSYSTEM'S QUALITY OF LIFE

Any population—human, animal, or plant—affects the ecosystem, and an increasing density of any population increases that effect. Evidence of the negative impact of increasing human population density on the quality of life on the earth is growing dramatically. Already in the early 1970s, scientists were making direct connections between the population explosion and its negative effects on the quality of the air and the water and other natural conditions.[12] A consensus began to emerge that "the major demographic problem facing the world is not saving lives, but rather slowing down the rate at which people are being born."[13]

A recent compilation of the research on environmental issues concluded that the leading crises facing the earth are unmet human needs for safe water, food, shelter, health care, education, and employment; encroachment on wildlife habitat and species depletion; land and soil degradation; depletion of nonrenewable energy and minerals; depletion of fresh water; water and air pollution; depletion of the ozone layer; global warming; and conflict and war.[14] After analyzing all the evidence and research reports, the *Global Ecology Handbook* concluded that increasing population density was the basic cause for all of these problems and that "slowing population growth is the most effective single measure for alleviating the range of global problem impacts."[15]

Water: Availability and Quality

Nearly 71 percent of the earth's surface is covered with water, but only a small percentage (0.003%) is available for immediate human use.[16] Its unique properties—liquid form (0–100° C.), high heat capacity, high heat of vaporization, efficient solvency, expansion properties upon freezing—all make water a precious resource without which living things as we know them could not possibly survive. Still, only a fraction of Earth's water is available for human consumption. About 97 percent is salt water, and the remaining 3 percent is largely unavailable (ice caps, glaciers, deep deposits).

Fortunately, until the present, the natural water cycle has ensured that even the small amount available to humans is recycled and naturally purified. Still, increasing population growth, expanding industrialization, and global climatic shifts make any long-term positive outlook rather grim. At issue are increasingly disparate local needs (especially around large urban centers), the "mining" of underground water supplies faster than they are replenished, and the overloading of renewable water supplies with too many wastes.

The fresh water needed for human use comes from surface and ground-water supplies. There is at least forty times as much ground water as surface water. Some underground reservoirs (called *aquifers*) are huge, largely nonrenewable containers of precious water. Some aquifers underlie major deserts (the Sahara in Africa) and several large central continental regions (the Ogalala Aquifer in the Great Plains of the United States). Tapping water from these reservoirs amounts to water mining and poses a problem for long-continued use.

The United States has the highest per capita water withdrawal in the world, and, given present trends, future supplies of fresh water will become increasingly scarce, especially for irrigation and manufacturing uses. Many areas of the world now experience severe water shortages. Moreover, the long-range effects of drought become intensified by over-grazing, unsound farming practices (including serious salt buildup from irrigation), deforestation, and the like. If global warming increases, as predicted, over the next several centuries, drought-prone areas can only increase. As the need for more water increases, freshwater supplies will become a more serious political issue. Water's value as a commodity will increase.

Heavy rains and ensuing floods create their own problems and often do little to ease long-term demands for water. Industrialized countries generally have a problem with water supplies contaminated by various sediments, excess nutrients (especially in high-impact agricultural areas), pathogens, and various hazardous chemicals.[17] When over 1.5 billion persons lack safe drinking water and 1.7 billion persons lack acceptable sanitation facilities, the need for action becomes imperative.[18]

Atmosphere: Air Quality

Without wind and air circulation, the earth would be virtually uninhabitable. Most of the atmosphere consists of two gases: inert nitrogen (78%) and oxygen (21%).[19] Several other gases, including carbon dioxide (0.036%), make up the remainder. Throughout its history the composition of Earth's atmosphere has varied considerably. Human activities are now adding to the complex mix of gases that makes up the atmosphere, which consists of four layers: the troposphere (extending about 17 km, or 11 miles, above sea level), the stratosphere (17–48 km, or 11–30 miles, above sea level), the mesosphere (a region of scattered ozone molecules so important in screening dangerous ultraviolet radiation), and finally the thermosphere (a region of rising temperatures due to molecular bombardment of atmospheric molecules by solar and cosmic radiation).

Global air circulation in the troposphere is caused by varying amounts of solar energy and uneven heating of the earth, reinforced by its axial tilt and rotation, resulting in large convection cells. With the peculiar properties of both air and water—the different amounts of air that water can contain at different temperatures, the different amounts of water that air can hold at different temperatures—we are swept with a variety of currents, most of them unpredictable.

One of the concerns facing humans in the twenty-first century is the so-called greenhouse gases (CO_2, water vapor, trace amounts of ozone, methane, nitrous oxides, chlorofluorocarbons [CFCs], etc.), which block the escape of infrared radiation and consequently warm the troposphere. The interplay between the ozone in the stratosphere and water vapor in the troposphere serves to buffer extreme swings in temperature. The loss of ozone coupled with an increase in various gases in the troposphere may cause serious, life-threatening climatic changes.

Although some people doubt the "greenhouse" threat, various danger signals are emerging:

—A steady two-decade increase in the major greenhouse gases (CO_2, CFCs, CH_4, N_2O) due to human industrial activities

—A rising mean sea level due especially to the warming of Antarctica and Siberia

—Increased storm activity and unusual weather, especially in the midcontinents and midoceans

—The retreat of glaciers

—Increases in tropical diseases, both human and vegetative

Although the mean surface temperature of the earth is expected to rise (1.8–6.3° F.) with certain climatic changes, no current model can unequivocably predict major climatic events in 2050.

Even with only moderate global warming, possible effects will be noted in food production (especially if increasing desertification occurs in the continental land masses in the Northern Hemisphere), shifting zones of forestation/deforestation, and rising sea levels. These moderate effects will have an immoderate influence on human health, global food supplies, and weather patterns. It follows that an altered lifestyle must be undertaken and soon. Basically, we need to eliminate all CFCs, cut the use of fossil fuels, use energy more efficiently, switch to sustainable agricultural practices, halt deforestation, and stop population growth.

Pollution: Waste and Health

All forms of organic life must have available energy (a ready supply of food) and be able to eliminate various organic wastes. Despite high-tech living styles, all wastes go somewhere, and many are recycled back into our own bodies. Most wastes today are manufacturing by-products (e.g., sulfur dioxide emissions from coal-fed steam generating plants) or pesticides designed to control insect infestations. Pollution can take many forms. Some pollutants, like smog, are obvious, while others are more subtle—salt overused on roads, fertilizer run-off from fields, and the like.

Physical hazards. All of us are exposed to various forms of ionizing radiation: x-rays, radioactive isotopes, radon-222, ultraviolet radiation from the sun, background radiation (cosmic rays, etc.), neutrinos from nuclear fission/fusion reactions. Such ionizing radiation poses a hazard by damaging cells in human tissues, causing either genetic damage due to altered chromosomes and genes or somatic damage, such as various cancers or cataracts. How much radiation a person can be exposed to without risk is subject to debate. Nevertheless, over a short time ionizing radiation in large doses can be fatal.[20]

Chemical hazards. The industrial revolution has markedly altered the lives of all humans. Toxic chemicals are everywhere and can play a significant role in human health. Many chemicals, even at low doses, are mutagenic (i.e., can cause changes in the genetic material). Other chemicals can cause defects in the growth of human embryos. About a thousand new chemicals are introduced each year, but many have never been screened for their toxicity to humans. Yet the increased use of chemicals by larger populations is part of the ecosystem in which humans live.

Biological hazards. Increasing population densities compounded by increasing poverty—overcrowding, unsafe water, malnutrition—spell disaster due to diseases, some transmissible (viruses, bacteria, parasitic worms)[21] and some nontransmissible (respiratory ailments, cancers, diabetes, cardiovascular problems). One of the keys to a sustainable future for the species of the earth will be the ability of humans to recycle life-supporting raw materials such as water, air, and the various chemicals needed for good health. All wastes go somewhere and, as Garrett Hardin wrote over thirty years ago, the present human predicament is that the commons—our shared resources—are now increasingly misused or polluted.[22] Whenever larger numbers of people heedlessly dump pollutants into the atmosphere, assuming that dispersal will take care of the problem, degradation increases. When population density was far less than it is now, such pollution did not greatly matter.

Cultural hazards. People can pollute or damage their bodies through choices that create unsafe living conditions. Poor diet, smoking, drug abuse, binge drinking, unsafe sex, crowded housing, piles of garbage, inadequate disposal of human excrement, poor sanitation—all contribute to diseased bodies and unhealthful environmental conditions. Too many persons, despite better education and appropriate warnings, continue to persist in habits not conducive to good health.

There are also social consequences of degradation of the environment: residential and urban congestion, traffic and industrial noise, transportation gridlock, the storage and cost issues involved in waste disposal, and general pollution caused by overcrowding. We agree with the position now increasingly taken by the scientific community that high population density (i.e., overcrowding) is *the* most important independent variable causing degeneration of the physical environment and the quality of life for all species. Population growth and overcrowding not only exacerbate the negative effects of congestion and deny the continued existence of a natural habitat for wildlife, but also negate even such simple pleasures as solitude and clean, fresh air. A sustainable world, which is the only responsible and ultimate solution to the environmental crisis, is clearly dependent on stabilizing the world population, for only humans have produced wastes that do not become nutrients for the continuing ecosystem.[23]

These negative consequences of overcrowding do not mean that a given population density is absolutely good or bad; rather, in every dimension of human existence there is a limit beyond which increased numbers of people will threaten the viability and quality of life for all flora and fauna.[24] At the present rate of global despoliation, achieving a stabilized population is the most important immediate goal if the planet is to be saved.[25]

THE IDEOLOGIES OF POPULATION GROWTH

Religious Values and Reproduction of the Species

Most world religions, as well as localized tribal religions, have accepted propagation of the species as an unmitigated good—a part of the natural order. These religious systems have espoused a belief and symbol system that accepts the natural processes of birth, life, and eventually death as a part of the "order of things" needing no mysterious or epistemological explanation—the perpetuation of the human species has been taken as a given. Children are the means by which this perpetuation is secured.[26]

Religions generally have promoted a pro-fertility stance.[27] Beyond this basic orientation, specific pronouncements about the status of children, how many, their treatment, and their relationship to parents, other family members, and the larger society are infinitely varied. Religions have almost universally rejected infanticide and abortion and promoted the nurture and protection of the young in the ways of the faith. Above all, religions generally have considered large families with many children a desirable goal—reflecting the blessings of the higher power and of a like-minded society.

In the Judeo-Christian tradition large families have been taught as the norm, illustrated by the adage, "Lo, children are an heritage of the Lord, and happy is the man that hath his quiver full of them" (Ps. 123:3–5 AV). Religious values regarding the bearing of offspring have been derived from "natural law" arguments—that is, according to the "law of nature," living species demand offspring to perpetuate themselves as a species. Thomas Aquinas reinforces the "naturalist" interpretation: "It is fitting that the multitude of individuals should be the direct purpose of nature, or rather of the Author of nature. Therefore to provide for the multiplication of the human race, He established the begetting of offspring even in the state of innocence."[28]

Societal Values and the Production of Children

The purpose and role of offspring can be argued from the basic social and institutional structures that emerged to guarantee the survival of society itself.[29] In this perspective marriage and the family have evolved and survived to serve the procreation, protection, and nurture of children as well as their economic survival, including the control, allocation, and transmission of property. Kingsley Davis proposes that the family serves society by the "reproduction of new individuals, the nourishment and maintenance of these individuals' [economic functions], and the placement of these individuals in the system of social position."[30]

Obversely, the actions of the individual help create the economic, social, and psychological matrix that assists the formation of the family and its perpetuation. Every child is a form of economic and social security. This function is widely assumed to be the reason that large numbers of children have been the norm in most traditional and preindustrial societies: "High fertility was normative [and is] an adjustment both to high and variable mortality and to the central importance in community life of familial and kinship ties."[31]

If children serve economic and social needs, then it is logical to assume that more children are better.[32] There are, of course, many mitigating forces—physical, economic, medical, psychological, social, and religious factors—that have kept families from having the optimum number of children. Some societies, for example, have limited the number of children for economic reasons. The Tikopia, "keenly aware of the food limitations of the island, consciously attempt to limit the number of children. The Ariki Tafua insists that a man shall have only one daughter and one son."[33] According to Davis, "society limits fertility by its institutions and customs, [but] it also encourages fertility within the limits of these institutions and customs."[34]

Surveys among families of varied societal groups reveal that ideal family size is highly variable and, of even more importance, in every society the ideal family size is far above the replacement level obtaining at the time. Davis suggests that "zero population growth is unacceptable to most nations and to most religious and ethnic communities."[35]

Many demographers have proposed that the "demographic transition"—the theory that birth rates go down as economic conditions improve—will lead to a stable population base.[36] However, other research proposes that smaller family size is not based on economic factors alone, but on values that at least promote growth. As a consequence the reproductive rate has always been at maximum levels for the individual family within the limits placed by the prevailing conditions. Yet what each family has considered optimum turns out to be destructive of the larger good. Garrett Hardin describes it in his classic "tragedy of the commons:"

Picture a pasture open to all. It is to be expected that each herdsman will try to keep as many cattle as possible on the commons. Such an arrangement may work reasonably satisfactorily for centuries because tribal wars, poaching, and disease keep the numbers of both man and beast well below

the carrying capacity of the land. Finally, however comes the day of reck-
oning, that is, the day when the long-desired goal of social stability becomes
a reality. [The tragedy]: each man is locked into a system that compels him
to increase his herd in a world that is limited. Ruin is the destination toward which all herdsmen rush, each pursuing his own interests. Freedom
in a commons brings ruin to all.[37]

From a family perspective, another child, assumed to be beneficial in
effect, contributes to a collective disaster: Maximizing the number of
children in a family ultimately works against its own best interests. This
"survival" function of children has caused demographers to assume that,
as soon as the economic survival role of children is displaced by technology or increased economic prosperity, surplus births will disappear.
Families of four and five children in affluent suburbs in today's America
seriously challenge this assumption. Almost all research on family procreation indicates that there is little connection between family motives
for having children and considerations of societal or ecological well-
being. Religious and social values promoting procreation for personal
benefits tend to ignore the larger consequences of childbearing.

National and Political Values and Childbearing

National values and policies typically derive from and reflect religious and social values. In addition, nation-states have normally encouraged the production of children to serve national goals, including
the fighting of wars. Thus, Plato promoted the production of a superior
class of people whose role it would be to conduct the wars of defense and
expansion: "Their bravery will be a reason, and such fathers ought to
have as many sons as possible."[38] Similarly, Nazi Germany promoted
large Aryan families as a means of achieving national policies.

An indication of the positive national stance toward childbearing is
the resistance most nations have shown toward initiating programs to
control births.[39] Aided by the high values families have placed on the
right to have as many children as desired, nations have generally hesi-
tated to institute birth control policies. This has been true especially
in underdeveloped countries, which have suspected a genocidal objective on the part of the powerful, developed Western countries when they

have promoted family planning through the United Nations or other forums.

Economic and Market Mechanisms and Population

Political and economic systems and populations are highly interdependent phenomena in every society. In the Western world, the self-sufficient family/village subsistence economy has evolved via the capitalist/market industrial system into the so-called postindustrial consumer age—truly one of the most amazing phenomena in the history of civilization. This evolutionary process involved the interplay of an increasing population, an exponentially burgeoning technology, a changing social organization, and an increasingly secular value system,[40] which produced a system demanding ever-increasing markets to justify ever-increasing production. "The ideology of economic growth has been intimately connected with expanded and concentrated capital and production for much of the nineteenth and twentieth centur[ies]."[41]

In *Captains of Consciousness: Advertising and the Social Roots of the Consumer Culture*, Stuart Ewen analyzes the history of consumerism in America: "The mechanism of mass production could not function unless markets became more dynamic, growing horizontally (nationally), vertically (into social classes not previously consumers) and ideologically. Now men and women had to be habituated to respond to the demands of the productive machinery." Anthropologist Jules Henry explains that markets have to be created in order to sell the products industry mass-produces. "Advertising methods are related to a first tenet of American business: profits must increase without limit." He continues, "In contemporary America children must be trained [and created?] to insatiable consumption."[42]

A tacit understanding, therefore, has guided market economies in both the short term and the longer run—more babies create more consumers: "The Baby Biz: From Diapers to Day Care, Babies Are Big Business" (*Washington Post*, 4 October 1996). Thus, the production and distribution cycle from baby diapers to teenage entertainment to real estate has been dependent upon growth in the population. The ideological consequences are clear—maximization of profits depends upon increased production, which demands increased consumption, which implies more consumers, more babies. Jules Henry puts it succinctly: "The increase in the population of the United States and a rising living standard during nearly a century of rapid growth of productive facilities have helped solve the problem of the spiraling relationship between production and the need for an expanding market."[43]

The cultural ideology expressed in the production-consumption-profits spiral permeates and informs our Western industrial society. As early as the 1970s, Ehrlich and Holdrin suggested that "the decision for population control will be opposed by growth minded economists and businessmen, by nationalistic statesmen, by religious leaders, and by the myopic and well-fed of every description."[44] The connection between population growth and the production/consumption/profits system has a time-honored place in the "land, labor, and capital" mantra. Clearly, the supply of labor is a signally important factor in the functioning of the modern market economy. But so are markets, and both labor and markets assume babies. Recent treatments of the role of women in society argue that reproductive activities "are productive in their own right and should be valorized, claiming their share of social and economic power." But the function of child rearing, an imperative for the economic domain, is now seen from the feminist perspective as the "underlying source of economic exploitation, domination and male supremacy."[45] As Bratton argues, "One of the keys to population justice [birth reduction] is justice for women."[46]

The reasons for the reluctance to face the issue of high birth rates (sine qua non for the "growth society") should be obvious, and the implications for environmental degradation are easily predicted. To put the issue bluntly: No one dare question the interdependence of economic prosperity and more babies, for we all benefit.[47] In the language of our market economy, more babies mean more consumption, and more marriages mean more housing starts, which mean more jobs, ad infinitum. Not surprisingly, housing starts are the basic governmental statistic used to measure the health of the growing economy.

THE INTELLECTUAL AND PHILOSOPHICAL TRADITION REGARDING POPULATION GROWTH

The various philosophical traditions have almost invariably supported population growth, but this position has rarely been analyzed.

One reason for support is the biological nature of reproduction. Insofar as reproduction is seen as a natural phenomenon, like breathing and eating, there is not much reason to develop a philosophy of having children. Humanity has produced philosophical systems analyzing natural and biological reality but little on the purpose of the individual organism. Since there is little philosophical discussion of the *purpose* of a human child, there is little reason to expect a philosophy dealing with the appropriate quantity of children. Further, the relative inability of parents throughout history to produce an optimum number of offspring makes a philosophical approach rather superfluous. If the deities, fate, disease, and other uncontrollable factors determine the number of children in a family, it is rather academic to expect the emergence of philosophies of how many children a family should have.[48]

Maximizing family size has had practical support, derived from economic and social needs, but limiting family size has received very little philosophical emphasis. This supports the argument that a philosophy does not normally concern itself with issues that cannot be achieved. If the procreation process cannot be controlled, why talk about desirable limits?

The assumption that human society is worthy of propagation and expansion is expressed in the famous adage, "Be fruitful and multiply and replenish the earth." In the Christian tradition this is premised on the idea that creation (including reproduction) is good. Until the idea that there are limits to the earth and to growth began to appear, it was not necessary to think creatively about restricting growth.[49]

The idea of increasing the good (i.e., the greatest good for the greatest number of people) as expressed in the utilitarian philosophy includes a tacit assumption that, if X number of people pursuing the good life is good, more people doing the same thing must be equally good or better. There is little evidence that economists, demographers, politicians, business leaders, capitalists, and even philosophers have assumed anything other than the belief that more roads, more developments and more houses, more production of consumption goods, more consumption, more profits, more choices, and, underlying all this, more people are better.[50]

The most specific philosophical orientation underlying the "more is better" approach is "the unidirectional causal paradigm" of Western culture.[51] This is the assumption that there is a one-way flow of influence from cause to effect, that the state of a system at any given time is uninhibited by the state of the system at a previous time. Thus, an event in the past does not affect a future choice. Until recently, there were no intimations that past actions could limit or predetermine the nature of choices in the present reality or future trajectory.

THE EMERGENCE OF THE IDEA OF LIMITS

A strong philosophical basis for believing that there are limits to most classes of physical, biological, and social reality has only slowly emerged. The vastness of the physical universe, now understood as ever expanding, clearly does not suggest that there are cosmic or cognitive/ social limits. The evolutionary perspective contributed to the rejection of the idea of constraints or limits.

Humans have always accepted lesser limits for many elements of human and cosmic reality—for example, that humans lived "three score and ten,"[52] or that a horse can run only so fast, or that two and two more equal only four. But the warning of an ultimate incompatibility among a growth-oriented ideology, its supporting social system, and the environment has been raised by social observers from economists to sociologists, such as Robert Heilbronner, Karl Polanyi, Hazel Henderson, E. F. Schumacher, and Daniel Bell. According to Henderson, "It was only about a decade ago that the public became aware that the environment was being strained by industrial society. For the first time, there is discussion of whether more economic growth is sustainable."[53]

In chapter 1 the Harders present the material for this discussion. Martin O'Connor bluntly asks, "Is capitalism sustainable?" He answers no and continues that, "although capitalism on a world scale may be viable for some time yet to come, the prices to be paid in terms of both human conflicts and lost human, cultural, and ecological riches will be very great." Others are proposing that we are witnessing the breakdown and demise of the industrial society.[54]

Our awareness of the greenhouse effect developed only after smog and air pollution literally became visible and dangerous, illustrating the principle of feedback: The consequences of human actions become noticeable after the limits of nature's ability to "heal" have been approached or surpassed. Further, the reality of pollution in industrial society, illustrated by the production of carbon dioxide, especially by the internal combustion engine, became known only after sensitive instruments were developed in 1958.[55]

The publication of *Silent Spring* by Rachel Carson in 1962 and the emergence of the science of cybernetics (feedback) began to awaken us to the fact that there are unknown or unintended consequences of our actions. When our actions began to reach dangerous proportions, the feedback loops began to tell us that our actions were working back on, or changing, the system itself.[56] This introduced what Henderson calls the morphogenetic syndrome, that is, "structural transformation not inferable from the existing state of a system or its properties or variables."[57]

The parable of the "tragedy of the commons" tells us that the universe has built-in limits to everything, Western philosophy and ideologies notwithstanding. There *is* a limit to how many people can live on Earth and have a "good" life; there *is* a limit to how many people the earth's soil can feed; there *is* a limit to how much "development" there can be before the natural ecosystem is irrevocably altered or even destroyed;[58] there *is* a limit to how many people can live on a finite globe before the natural habitat for wildlife will be irrevocably damaged and destroyed.[59]

But human societies have always tried to avoid the idea of limitations, rejecting the fact that doing what seems beneficial for each person individually often results in future collective destruction. Solzhenitzyn moves straight to the heart of the ideological basis of our "unlimited" philosophy: "As far as I know, no state has ever carried through a deliberate policy of self-limitation ... The concept of unlimited freedom is closely connected in its origin with the concept of infinite progress, which we now recognize as false. Progress in this sense is impossible on our earth with its limited surface area and resources."[60] After describing how both socialism and capitalism have plundered the globe, Solzhenitzyn writes, "Either we change our ways and abandon our destructively greedy pursuit of progress, or else in the twenty-first century, we will perish as a result of the total exhaustion, barrenness and pollution of the planet." There is no more powerful and succinct way of stating the crisis. Solzhenitzyn offers a two-part solution to this critical problem— repentance of our rapacity and application of the idea of limitations on all of our actions. "With the population rapidly soaring, mother earth herself will shortly force us to [limit population growth]. It would be spiritually so much more valuable, and psychologically so much easier, to adopt the principle of self-limitation—and to achieve it through prudent self-restriction."[61]

Reality or a world-view, however it be defined, is derived from a philosophical and religious basis, and limits are almost universally implied, if not stated. But they have been avoided. Judaism and Christianity introduced a system of morals and ethics that placed a vast number of restrictions and limitations on most human actions. These limits historically were often lamented and circumvented.[62] Most religions have developed systems of constraining human behavior—at their very crudest, the idea of revenge or reciprocity (an eye for an eye). No person is free to do anything without running finally against the limitations imposed by nature. But this principle has been late to be applied to humans' relationship with the earth. Only after the truth of the principle of limits becomes a philosophical/moral/spiritual factor can it become a matter of public policy and application. Ironically, we are coming to this in exactly the reverse order—the harsh truth of the reality of limits in the physical world is causing us to come to the philosophical/moral/religious realization that a principle underlies the material universe. This should tell us plainly that we humans have been naive regarding our place in the universe.[63]

THE CHALLENGE BEFORE US

When and if a philosophical/moral/religious awakening to the idea of limits does come, it will be easier to implement a social and public policy for a sustainable environment. Realizing that limitations undergird all of creation, including human reproduction, will make it much easier to develop a social, national, and international policy regarding the control of population growth before brutal reality does it for us. Realization of the overpopulation crisis will require a massive shift in religious dogma and traditional teachings in North America, especially for the Roman Catholic and conservative Protestant churches, not to mention conservative marginal groups such as the Hutterites and Mennonites.[64]

The shift in religious and moral values toward controlling and limiting human reproduction will come; religions have changed many times in the face of hard reality. The only position that reasonable and responsible people can take is to adopt the foundational premise of limits and then to work toward a philosophy and policy of an optimum and sustainable population. Clearly, an almost universally accepted axiom is that a balance must be achieved between humanity and other parts of nature, a balance that will maximize the quality of existence for all forms of life. As Lester Brown puts it, "Awareness that rapid population growth is a threat to improvements in the human condition is now widespread."[65]

Many quality-of-life indicators, including zero population growth assessments, suggest that we have long passed the optimum population mark. Although a goal of negative growth is only a distant possibility, ultimately humankind will have to reduce its numbers to allow a sustainable and decent quality of life. If the principle of limits operates, there is obviously a limit, and if we have already passed it, as some people maintain, a reduction of population will be required to recover a sustainable equilibrium. In practice, this means that we all must give increased support to education on the overpopulation crisis. It means increased support of family planning and birth control programs and, above all, improvement of the status of women all over the world, not only in the lesser developed nations.

The challenge before us now is to restore the balance of human population with the rest of nature, which will help achieve the optimum life for all. We offer some concrete solutions in chapter 13. Here we focus on the central issue: Limiting the number of persons on the earth is the most effective means to achieve the goal of preserving (salvaging?) a sustainable, equitable, and "high-quality" natural environment. "Slowing population growth is the most effective single measure for alleviating the range of global problem impacts."[66] The final report of the Twenty-third American Assembly on the Population Dilemma concluded that "the vast majority of the world do not yet recognize the full implications of present population trends ... The Twenty-third American Assembly cannot emphasize too strongly that time is running out ... To continue to ignore world population problems is to ignore the welfare and security of all peoples. We must not remain complacent in the face of a major threat to world peace and survival."[67]

We enter the twenty-first century acting on the right of unlimited growth and procreation, a sacred cow derived from the prevailing ideology that human beings have been placed on Earth to pursue their unlimited ends without external limits. Can humankind be saved from this folly before it is too late?

PART II

Anabaptist/Mennonite Life & the Environment

CHAPTER 4

God's Spirit and a Theology for Living

David Kline

The most commonly used prayer book in Amish homes is *Die Ernst-hafte Christenpflicht* (The Serious Christian Duty). It dates to the early 1700s and the Palatinate Mennonites and also was common in many Mennonite homes, especially Swiss/German, until the Great Awakening of the late 1800s. Among prayers for many occasions, the small volume contains several evening (*Abend*) prayers.

I have a small, slightly condensed edition of the *Christenpflicht*. It was printed during 1826 in Wooster, Ohio, by Johann Sala. My parents got it when my grandparents' estate was divided among the heirs back in the early 1950s. On page 10 is the evening prayer "Ein schön Abend-Gebet täglich zusprechen," translated as "A nice evening-prayer to be read daily." The lower outside edges of those pages are so thumb-worn that the print is hard to read; the pages were illuminated countless times by the light of a candle or kerosene lamp.

Toward the end of the five-page prayer is the line "und lasz uns deine Creaturen und Geschöpf nicht verderben sondern dasz wir zur ewigen Seligkeit mögen gebracht und erhalten werden." Literally, this translates as "and help us not to harm your creatures and creation but that we may be brought to eternal salvation and may abide therein." I like a friend's translation: "and help us be gentle with your creatures and handiwork so that we may abide in your eternal salvation and continue to be held in the hollow of your hand." There, I think, lies the Anabaptist theology for living. Maybe, as I recently read somewhere, a man travels the world over in search of what he needs and returns home to find it. That line of the evening prayer swirls in a wispy cloud of controversy, of course. Good Anabaptists would not pray such an earth-centered prayer, would they? After all, our salvation does not hinge on how we care for creation or on what we claim as our profession.

To confuse the issue further, later editions of the *Christenpflicht* have commas inserted "und lasz uns deine Creaturen und Geschöpf nicht verderben," which shifts the focus of the prayer to us humans as God's creatures and handiwork and away from the rest of God's creation. Maybe our local Wooster printer Johann plucked out the commas on his own initiative. The few theologian-historians I asked said the commas should be included. The point is that hundreds of fathers and mothers belonging to the Anabaptist branch of Christendom did pray and believe exactly as the line of prayer reads in the Wooster edition of the prayer book.

A THEOLOGY OF LIVING FOR TODAY'S ANABAPTISTS

Why is it that some of today's Anabaptists are having such a difficult time finding a theology for living? Could it be that we have become too alienated from the land to which our foreparents felt so closely connected and simply can no longer relate to the creation? Creation has become something distant, something you go to, but has ceased to be a living part of us. Are we merely residing on the land instead of living with it? If we would believe that line of prayer without the commas, we would take care of God's creation because our lives depend on it. We would believe that "the earth is the Lord's and the fullness thereof."

We are friends of a young family whose farm was threatened by a housing development. The houses would have shattered the sanctity and privacy of their farm. When my friend told me what might happen, he had tears in his eyes. "I will weep if I leave, and I will weep if I stay. But I have to stay; the farm has been too good to us. I can't turn my back to these fields and pastures and trees [we were looking across his eighty-acre farm] that have nurtured and shaded our animals. In turn the animals have nurtured us, which has given us a good life." That young family needs no theology for living; they are living it. (Fortunately the housing plans fell through.)

The Mennonites and Amish have a long history of being stewards of the land. When the Palatinate Mennonite David Mellinger was asked in the 1700s about his success as a clover farmer, he replied, "I should have already given princes and other great lords a description of my operation and how I achieved it but I cannot tell it so easily. One thing leads to another. It is like a clockwork where one wheel grabs hold of another and then the work continues without my even being able to know or describe how I brought the machine into gear."[1] Louis Bromfield, novelist and later an innovative farmer on Malabar Farm near Mansfield, Ohio, wrote in his delightful book *Pleasant Valley* (1940) that the Amish and Mennonites are the only farmers in America who have stayed on the land they settled and have kept improving it.

Sadly, many Anabaptist families proved that "a man standing in his own field is unable to see it," as Emerson is thought to have said. They looked longingly at the industrial society and, to use an Amish phrase, jumped the fence from an agrarian life to an industrial one. Here is when the tendency to become an exploiter instead of a nurturer had its beginnings. When that link to the land or the earth is severed, life revolves around plastic, asphalt, steel on rubber, false-security lights—human-created things—and the weather becomes something to complain about or escape from. The beauty of the changing seasons is of minor significance. Nature becomes an adversary, something to be subdued and altered to one's liking, a resource from which to profit, seldom loved for its own sake.

Early Anabaptists looked at the earth not as an adversary but as a friend. Even bad weather had its advantages. When the weather was too miserable for the Taeufer Jaeger (Anabaptist hunters) to venture out, the believers had an opportunity to assemble for worship. Caves and forests sheltered them for secret meetings. The Jura Mountains became a sanctuary for those fleeing persecution in the Emmenthal. Perhaps that reliance on nature's protection is one of the reasons Anabaptists never much cared for purging from their everyday lives what today would be considered pagan practices. When pagan Europe was Christianized, great efforts were made to eradicate every practice of pagan or earth worship to create a distinct division between Christian and pagan practices. Since Anabaptist forebears were so rural and closely tied to the land that gave them sustenance, some "earthy" practices survived in their lives and rituals.[2]

The Anabaptist register of Scriptures (all New Testament) and hymns (from the Ausbund) tends to be seasonal, especially the hymns. In the spring we sing of the skylark trilling its love song and in the autumn of the coming cold season. Farm families prefer an autumn wedding because much of the food for the wedding dinner can be raised or grown during spring and summer—the corn and potatoes; stock for the dressing; apples for the pies; strawberries for jam; lettuce, broccoli, and cauliflower for salads; and broilers for fried chicken. The Old Order Amish still have their weddings on Thursdays and sometimes Tuesdays. (The more liberal Amish have switched from Thursday to Saturday, echoing the trend from agrarian to industrial.)

According to William Schreiber, "The Amish have preserved for modern society the clear traces of ancient pagan cults which pre-date the Christian era; cults based upon the worship of the pre-Christian Germanic gods Ziu and Donar (Donnerstag–Thursday). In this more than in any other phase of their distinctive life, the Amish form a cultural enclave stronger and more cohesive than any other known within modern America." Thursday is still the preferred day for weddings in parts of southern Germany. In rural South Germany one still heard in 1962, *Donnerstagheirat–Glücksheirat* (Thursday wedding–lucky wedding). "Here (South Germany) the god Donar was the preeminently revered god of marriage and of course it must be added that he is the god of agriculture, of the livestock and of fertile growth. Practically all farm life and increase was his special domain."[3]

Ascension Day, which always comes on a Thursday, is observed by the Amish as a day of fasting and prayer, a holy day. Interestingly, the nation of Germany observes Ascension Day in the same manner. Stores are closed and no trucks are allowed to travel on that day. The early Anabaptists (and now the Amish) were an earth-bound people who showed a reverence for creation and the earth. They never felt the angst of modern Christianity's separation from nature. As our young son once asked, "If Heaven is such a beautiful place and the earth so bad, why are people so reluctant to let go of life?"

Some scholars claim that the hymns sung by the Amish in church are simply the melodies of rural folk songs from the sixteenth century adapted to the Ausbund hymnal. Many of these hymns were written by believers in prison. Musicologists think that the hymns are also connected to Gregorian chants.

And I would guess that more Amish than will admit to it still plant some of their seeds according to certain astrological signs. My father would sow the spring clover in the sign of Leo in March. Some neighbors plant their potatoes on or the day after the May full moon, radishes in a waning moon, and above-ground crops in a waxing moon. Fence posts were dug in during a waning moon to prevent heaving and to ensure that there would be enough soil to tamp in and fill around the post. Dousing, or water "witching," is still a fairly common practice in spite of accusations of witchcraft. *Brauching*, which somewhat crudely translates as powwowing, used to be commonly practiced by a small number of healers in the community.

As a child I developed a severe case of hives. Late one evening my parents took me to a local *Braucher*, who moved his hands over the affected parts of my anatomy, all the while saying something I could not understand. What I did understand was that the next morning my miserable hives were gone. A story is told about how Dr. Hostetler came down with erysipelas, a condition his own medications could not cure. One night, like Nicodemas to Jesus, Dr. Hostetler walked the mile across the fields to Brauch Jake, who cured him with one treatment. No one argues with success. Why, as a friend asked recently, would an evil spirit want to heal? I have struggled not too successfully for a long time with Western culture's world-view that alienates itself from nature. The earth is perceived as only a stage on which to work out our destinies and the creation—trees, plants, insects, mammals, even the constellations—as merely props in the drama. We say that nature exists solely for our benefit, a vast supply warehouse to let us live sumptuously without regard for tomorrow.

Another point that has me puzzled is why some Christians argue so vehemently against the theory of evolution and then abuse the God-created earth as if they have a God-given right to do so? They seem not to have any qualms about being part of the systematic destruction of the natural world for the sake of human greed. There is much to be lost by such arrogance and lack of reverence for creation and its Creator. Recently, I was given a summarized copy of Dr. W. C. Lowdermilk's 1939 timeless report on the historic loss of twenty-seven of the world's civilizations. Civilizations and their great cities collapsed because the land that nurtured them was abused and depleted to the point of infertility. Babylon became, as the Hebrew prophets had warned, "a desolation, a dry land . . . and wolves shall cry in their castles."[4] Ironically, when Dr. Lowdermilk visited Babylon in the late 1930s, the only living thing he saw in the desolated city was a lean and rangy wolf. He viewed the ruins of Carthage and Timgad and the north coast of Africa, at one time the "granary of Rome," its hillsides bare of topsoil, even though the valley floors are still cultivated. Many of the ancient cities are buried by soil eroded from surrounding slopes, soil that at one time had made the cities prosperous. Out of Dr. Lowdermilk's shocking report on the devastation early civilizations suffered from lack of land stewardship, the U.S. Soil Conservation Service was born—a noble attempt to reverse the direction in which our nation was heading.

We say we know better now. We have learned from history. We have better technology with which to combat the forces that ravage the land. But do we? To save the soil we rely on disturbing amounts of chemicals and pesticides. Recently, I helped identify and then draw blood samples from Amish people with Parkinson's disease in a study conducted by Ohio State University working jointly with the University of Miami (Florida) under a grant from the federal government. Interestingly, Miami, New Orleans, and Millersburg, Ohio, have what researchers call *clusters* of Parkinson's sufferers, levels much higher than in the rest of the nation. The theoretical cause in Ohio is pesticides. Almost all Ohio patients are men and were active farmers in the early 1950s, when the insecticide heptachlor, a chlorinated hydrocarbon closely related to DDT, was heavily used to control spittlebugs in hay fields. The experts advised farmers to use this new technology because it was effective and long-lasting. According to a professor and researcher at the Ohio Research and Development Center, even after forty years, chlorinated hydrocarbon residues are still in the soil and, when the conditions are right, the persistent chemical volatilizes out of the soil and is reabsorbed by the crops.

How does a person or a church go about finding a theology for living? An Amish bishop from Pennsylvania said, "We should conduct our lives as if Jesus would return today but take care of the land as if He would not be coming for a thousand years." The "eleventh commandment," which Dr. Lowdermilk gave in a talk on stewardship in Jerusalem in 1939, is interesting: Thou shalt inherit the holy earth as a faithful steward, conserving its resources and productivity from generation to generation. "Thou shalt safeguard thy fields from soil erosion, thy living waters from drying up, thy forests from desolation and protect thy hills from overgrazing by thy herds, that thy descendants may have abundance forever. If any shall fail in this stewardship of the land thy fruitful fields shall become sterile stony ground and wasting gullies, and thy descendants shall decrease and live in poverty or perish from the face of the earth."[5]

ONE MAN'S WAY OF CARING FOR CREATION

My father practiced what Dr. Lowdermilk preached. My parents married in 1929 and began working a "farmed out" property sold at a sheriff's sale. In his forty years of farming, my father did what Bernd Langin wrote that the Amish strive to do: "Land ought to be treated and developed so that parents can face future generations without shame for what they have done to the earth."[6] When he handed the operation of the farm over to my wife and me in 1968, my father had to feel no shame; it had been cared for lovingly. "When you send forth your spirit, they are created; and you renew the face of the earth" (Ps. 104:30 NRSV).

Father died in March 1993. That first spring and summer, I was unable to return to his grave, for reasons I can't really explain. In September I walked the mile across the fields and woods to my brother's farm, where Dad is buried. On the way over I picked up a red tail feather lost by a molting hawk, a feather that had soared high over our farm but now, too, had fallen to earth. Entering the small *Friedhof,* I knelt and "planted" the feather on the mound of fresh soil on Dad's grave. Then I rose and looked over the neighborhood where the red-tailed hawk hunted voles, screamed its shrill cry from high in the cloudless sky, and raised its young—a land of fields and woods loved by the hawk and its mate and loved, too, by my father.

Looking down into the valley of Salt Creek, I recalled the many times Dad and I had walked along its meandering course through the pastures. Sometimes fishing, sometimes mushrooming, sometimes returning from a coon hunt; the walk was always pleasurable. Then I raised my eyes and looked at the farmsteads. To the east and south there hardly was a set of buildings that my father did not have a hand in building. Many of the barns had timbers he had sawn on his little sawmill. He never missed a barn raising or work "frolic" in the neighborhood. He enjoyed the camaraderie, he loved to help, and his skills in carpentry were useful.

Dad gave more to the world and its inhabitants than he took. He became native to this place, living on and from the land. Toward the end of his life, he told me, "Eighty-seven. It sounds like a long time. But it wasn't; it was a short time."

I think it was so because he loved life—the creation and the joys of seeing and being a part of it all. He never saw the Pacific or the redwoods or the Grand Canyon or even the Mississippi, and I don't think he felt deprived for not having beheld those natural wonders because he "saw" so much around home. He could tell every kind of tree by looking at its bark at eye level. He knew the locations of the only persimmon and cucumber trees in the area. Likewise, Dad knew every raccoon den tree for miles around. He would show me where the coons would climb a smaller tree that leaned to the limbs of the larger den tree—he could tell by the claw scratches on the bark. He never knowingly cut down a tree that was home to a raccoon or squirrel. When a neighbor sold an old sugar maple that was a den tree, Dad found it almost sacrilegious. Of course, he didn't tell the neighbor his feelings. When working in the fields, Dad could tell at a glance whether a large, soaring bird was a red-tailed hawk (long broad wings and fanned tail) or a turkey vulture (longer and narrower wings held in a slight dihedral). He knew the slower, ganglier wingbeats of the northern harrier (marsh hawk, he called it) and the swift wingbeats of the speedy Cooper's hawk. He could tell the condition of the soil by studying the "weed" plants growing in the field. If sorrel (sour vines in the dialect) grew, he knew the field needed lime. Pesky quack grass indicated a calcium deficiency.

Dad absolutely abhorred fall plowing. Leaving that freshly plowed, bare soil exposed over winter to wind and water erosion was totally unthinkable. It was much harder on the field than taking off a crop, he insisted. My guess is that his aversion went deeper than concern over soil erosion—plowing in autumn's chill just doesn't sit right. Plowing should be done in the renewal of spring when one shucks off the coat of winter and rolls up shirt sleeves to the returning sun and Gulf-warmed wind, which bring with them the meadowlarks and pipits and vesper sparrows, ideal plowing companions. I can easily see why ancient cultures in northern climates worshipped the sun. We farmers in the spring almost do.

My father's thinking on caring for creation was influenced by a Mennonite schoolteacher he had from second through eighth grades, Clarence F. Zuercher. Forty years later Mr. Zuercher became my teacher in the second grade and taught me for seven years. Both Mr. Zuercher and my father were descended from Swiss immigrants, and both were farmers and loved the outdoors. I will not get into the creation-caring methods Mr. Zuercher used for teaching, but he never let the classroom interfere with his students' education. If one's livelihood comes from the earth—from the land, from creation on a sensible scale, where humans are a part of the unfolding of the seasons, experience the blessings of drought-ending rains, and see God's spirit in all creation—a theology for living should be as natural as the rainbow following a summer storm. And then we can pray, "Und lasz uns deine Creaturen und Geschopf nichtverderben" (And help us to walk gently on the earth and to love and nurture your creation and handiwork).

CHAPTER 5

Mennonites, Economics, and the Care of Creation

Michael L. Yoder

The animals complained to God that every place they went, people would occupy, forcing them to leave. God said to them, I have one place where no human being will survive, so go there and you will live in peace, and that was the (Paraguayan) Chaco . . . And then came the Mennonites.
—Heinrich Ratzlaff Epp

Just as I began to write this chapter, the current issue of the *Nature Conservancy* magazine, complete with the usual full-color photos of exotic species needing to be saved from destruction, was delivered to my home. The pictures were of capybaras, giant anteaters, and maned wolves. More interesting, however, were the pictures of maté-sipping Mennonite Heinrich Ratzlaff Epp, dubbed by the author "the Savior of the Green Hell."[1] Like his fellow Paraguayan Mennonites, Ratzlaff Epp is a descendant of the Russian Mennonite refugees who have tried to recreate a Paraguayan version of the once-prosperous Russian Mennonite commonwealth. Only a few hundred indigenous people lived in the Chaco before the Mennonites arrived. Mennonite refugees came there, like their ancestors did to the swamps of the Vistula in Prussia/Poland and to the steppes of South Russia, and have prospered in and "developed" an area seen by most as inhospitable. Their descendants produce cattle, various food crops, and most of Paraguay's milk.

In Paraguay as elsewhere, however, development has come at considerable environmental cost. As the Mennonites opened the Chaco to development and cleared its forests for cropland, much the same fate that in North America awaited the once-vast prairies and the herds of bison began to unfold, but in a Paraguayan variation on the theme. Forests disappeared, native species retreated, and other even less stewardly opportunists entered. Ratzlaff Epp recounts his distress at seeing hunting parties returning to Asuncion on weekends (on the Mennonite-built Trans-Chaco highway) with truckloads of dead animals. "There was nothing I could do. There was no Paraguayan law against such an act."[2]

Having prospered as a psychologist, Ratzlaff Epp reluctantly allowed himself to be persuaded to run for a seat in the Paraguayan legislature, representing fellow Mennonites in the Chaco, and he won. Today he tries to pass laws to prevent the wanton destruction of the animals of the Chaco. He has also helped persuade the Chaco's largest landowner, Peter Durksen, to set aside as land preserves two properties totaling twenty-five thousand acres. There is considerable irony here. The Mennonite "Savior" of the Chaco is trying to save what Mennonites themselves unintentionally had a hand in destroying, just as Mennonite Corn Belt and Wheat Belt farmers in North America had a hand in destroying the vast prairie and most of its varied flora and fauna in North America.

Economic life, or the production, distribution, and sale of goods and services, is traditionally divided into three sectors. The primary sector of the economy focuses on the production of food through agriculture. The secondary, or manufacturing, sector transforms raw materials taken from nature into manufactured goods. The services sector provides us with the services we have come to take for granted in modern societies.

THE PRIMARY ECONOMIC SECTOR: AGRICULTURE

Holy Disturbance

The computer prints it out,
and with trained financial eye
my banker scans the sheet.
"Keep costs low, sell products high,"
he says with a twinkle in his eye.

"Let the profits mount, Bob.
It's the bottom line you count."

Ignore the bottom line?
No way. Business isn't play.
The bottom line's reality.
To face it takes God's grace.
For a Holy Disturbance comes to play
where faith and profit interface.

—Robert Yoder, 1996

It is fitting to consider first the agricultural sector: (1) Agriculture is the most basic human economic activity. (2) It has provided a livelihood for more Mennonites throughout history than any other endeavor. (3) It is still the preferred occupation of the Amish, the Hutterites, and conservative Mennonite groups such as the Old Colony and Church of God in Christ Mennonites. (4) While progressive Mennonite groups have lately been abandoning farming for other occupations, this change is very recent. Even progressive Mennonites still are more likely to be farmers than are non-Mennonites, many of them still have an agricultural background, and many still prefer to live in rural areas.[3]

Mennonites and their Amish and Hutterite Anabaptist cousins have a long and distinguished track record in agriculture. They are often credited with introducing crop rotation and the use of natural fertilizers, both animal manure and legumes, into European agriculture, raising soil fertility and crop yields. In addition to being seen as honest and hardworking by their neighbors, they also gained a reputation for being excellent livestock farmers.[4] Transplanted Dutch and North German Mennonites drained the swamps of the Vistula in Prussia/Poland, making it a productive farming area, and then used the reclaimed land to produce excellent milk and cheese. Some of the same Mennonites were later invited by Catherine the Great to enter South Russia, where they developed extremely productive agricultural colonies.[5]

To some degree, necessity was probably the mother of invention for early European Mennonite agriculturalists. Often denied the chance to own land as a persecuted religious minority, Mennonites were forced to seek employment as tenants and managers on estates of large landowners. If they did have a chance to farm for themselves, it was often on marginal land, the hilly and less fertile land of Switzerland and Alsace disdained by other farmers or the swamps of the Vistula. Mennonites often had smaller farms than the average, at least in Switzerland, France, and South Germany.[6]

Intensive livestock management and production allowed Mennonites to make a living on these small farms, usually as tenant/managers. The animal manure thus produced was used as fertilizer, increasing soil fertility and crop yields. Hard work, intensive and honest management, and practical knowledge gained through experience combined to make these farmers productive and sought after as tenants, managers, and settlers in new areas. They learned to work with nature in an environmentally friendly way. Modern research-based scientific agriculture did not yet exist, nor did the modern environmental movement, of course.

These Mennonite agriculturalists brought their practical knowledge of farming with them to the New World. In the 1700s, 1800s, and even early 1900s, it is doubtful that Mennonite farming differed much from Amish farming. All farming was done by hand or with horses and oxen. Some soil erosion no doubt occurred when land was cleared, plowed, and planted to row crops, but the need for forage for horses and other animals necessitated keeping large amounts of land in pasture and hay and oats (used as a nurse crop to establish pasture and hay seedings). Forage and small-grain crops shielded the soil from the pounding of raindrops year-round, and roots also helped keep soil in place. Few chemicals were used on farms. Crop rotation and the use of animal manures kept fertility relatively high, as they had on European Mennonite farms. A variety of crops and animals were raised, as on Old McDonald's mythical farm, providing a measure of insurance against failure affecting only one crop or animal species in a given year. It was possible to make a decent living for oneself and one's family on horse-powered farms, not unlike today's Amish farms.

In the twentieth century, modern farming diverged radically from the earlier mode of agriculture still retained to large degree by the Amish. Tractors of increasing size do what horses, mules, and oxen did and do it much faster, making it possible for one farmer to farm 1,000–1,500 acres instead of only 80 or at most 160, as with animal power. Hybrid seeds, chemical fertilizers, and chemical pesticides and herbicides have all helped to increase crop yields significantly. Modern farmers, including most Mennonite farmers, have become gradually but increasingly dependent on all of these, mostly for economic reasons. Especially in the Corn Belt, livestock farming with associated soil-conserving forage and animal manure by-product has given way to intensive crop farming, partly for reasons of convenience, but mostly for economic reasons. An acre of corn simply produces more net return than livestock raised from forage crops (and with less total effort).[7] Where livestock are raised, especially hogs and poultry, it is often in confinement operations, which increase problems of odor and manure disposal.

The cost-price squeeze in North American agriculture is such that the costs of inputs used in modern farming (machine power, chemical fertilizers, pesticides, and herbicides) have all increased tremendously while the prices received by farmers for their agricultural products have not. New tractors used in Corn Belt farming cost up to $100,000, and combines cost up to $175,000.[8] The cost of hybrid seed and chemicals used intensively in corn production is hardly "chicken feed." Insurance is purchased by most farmers to guard against the possibility of hail, flood, or drought ruining the crop. If a farmer is heavily in debt to the local banker for purchased land, machinery, and crop inputs, one bad crop without insurance can mean bankruptcy. At the same time, prices paid for basic agricultural commodities have risen much less over time, mostly because of an oversupply of grain produced by a combination of high yields per acre and increased acres devoted to grain production.[9] With high costs of production and low commodity prices, under normal circumstances the profit margin per acre is quite low after all expenses are paid.

For modern North American farmers, including Mennonites, the pressure is to "get big or get out." Farmers can no longer treat farming simply as a way of life. They are forced to watch carefully "the bottom line," as captured in my father's poem quoted at the beginning of this section.[10] Farming has become a business, often a cutthroat business as farmers compete against each other to buy or rent more land, raising prices for both to uneconomic levels. Prime cropland in North Central Illinois now brings up to four thousand dollars per acre, and farmers are willing to pay up to two hundred dollars an acre per year to cash rent land. Some become overextended financially, or their operations become unprofitable. The result is that many go bankrupt, as happened with a vengeance during the Farm Crisis of the 1980s.[11] Big fish swallow little fish on the farm as well as in lakes and rivers, reducing the numbers of Mennonite as well as non-Mennonite farmers. The bottom line is often cruel.

The few farmers who are left, Mennonite or not, feel forced to farm big. As long as they wish to live a middle-class lifestyle with automobiles, air-conditioned homes, modern medical care, vacations far from home, computers to keep track of increasingly complex accounts and new information available on the Internet, a college education for children, and so forth, most feel that they have no alternative. Farmers, Mennonite as well as non-Mennonite, have gradually become dependent on the technology of the modern world.

Sadly, chemical-intensive crop farming is rarely environmentally friendly. Row crops expose the soil to erosion much more than do forage and small-grain crops. Tillage with heavy tractors compacts the soil, reducing soil tilth. Excess nitrogen fertilization raises nitrate levels in ground water, sometimes making water from shallow farm wells undrinkable. Excess phosphorus fertilizer run-off increases eutrophication of lakes. Heavy use of herbicides also pollutes ground water to levels often unsafe for human use and may reduce the soil's organic matter over time, thus reducing both fertility and water-holding capacity. Pesticides often kill desirable insects like earthworms along with the targeted pest. Where used on vegetable and fruit crops, as by the Mennonite farmers of California's Central Valley, they may endanger farmers and agricultural workers who tend the crops, as well as consumers who eat them.

Irrigation offers a chance to grow bountiful crops in arid regions like California and the western plains but increases soil salinity and rapidly draws down the water table, as in Nebraska, where the huge Ogalala Aquifer is rapidly being depleted. In the short run, the economic system and modern agriculture reward those who use chemicals and machine power and irrigation water. In the long run, the patrimony of future generations is threatened. Most Mennonites have joined their non-Mennonite neighbors in farming the modern way. They see few alternatives.

Though little information compares Mennonite farming practices with others, a recent research report is rather revealing. Data from a random sampling of Mennonite and non-Mennonite farmers in central

Ohio in 1991 showed that Mennonite farmers were less likely to use modern practices such as no-till farming, holding to their traditional practices. "Since Mennonite farmers value tradition and have observed that traditional practices [such as deep plowing] produce high levels of output, they continue to use the practices even when they are no longer recommended." In conclusion, the authors suggest that "Mennonite production systems are not as environmentally benign as many people believe their farming systems to be."[12]

There is some hope, however. Some modern Mennonite farmers are learning to reduce trips over the field, saving labor, fuel, machine expense, and soil tilth. Minimum tillage and even no-till farming also result in increased crop residues left on the soil to break the force of raindrops and wind, saving both soil and moisture. Some are also practicing more selective use of pesticides, using them only where absolutely necessary rather than indiscriminately, as in the recent past. Roger Kennell sees these practices both as good stewardship of the land and as good economics, since they can result in considerable cost savings per acre. He notes, however, that "I have not continued straight no-till. I was sacrificing production and profit. I couldn't stay in business long that way."[13] This admission by an environmentally sensitive Mennonite farmer indicates that it is, in fact, economic forces that often win out over desires to be environmentally "pure." The bottom line wins again.

The Shared Farming movement is another recent environmentally friendly innovation that has been promoted by some Mennonites, such as "Farmer Dan" Wiens of Manitoba.[14] Essentially, it consists of farmers contracting with a group of urban consumers who purchase shares in the farming operation in return for a market basket of organically grown vegetables weekly during the growing season, at least for a year at a time. Purchasers are often willing to pay a premium for the promise of organically grown produce, and some have shown considerable understanding when bad weather has reduced the selection and quantity of vegetables available, as happened in Manitoba when a cool, wet summer prevented the tomato crop from ripening.[15]

Such arrangements may make it possible for environmentally conscious organic farmers, Mennonite or not, to make a living on five to twenty acres, greatly reducing needed investments in land, machinery, and chemicals as compared to conventional farming. However, such arrangements are fairly new and probably limited in terms of the numbers of such farmers who can be thus employed. Most consumers still want the cheapest food, not the environmentally cleanest. Additionally, there is some evidence that total profits on most small-scale organic farms do not usually approach those of more conventional large-scale farms.[16] So economics again may limit the applicability and success of this promising option.

There remains, of course, the Amish option of going back to old-fashioned, low-input, horse-powered farming, with its crop rotation, natural fertilizers, and minimal use of environmentally harmful chemicals and nonrenewable energy in the form of fossil fuels.[17] Even there, however, some soil erosion occurs, as Amish farmers prefer clean tillage with plows and cultivators rather than the minimum-tillage practices of modern agriculture. And for most progressive Mennonites who wish a middle-class lifestyle, the Amish option is not really a possibility, as it will rarely produce enough income to make such a life possible. Even many of the Amish are being forced to turn to off-farm sources of income, working in factories or developing cottage industries.[18]

THE SECONDARY SECTOR: MANUFACTURING AND CONSTRUCTION

Compared to agriculture, Mennonite experience in manufacturing enterprises is much less extensive and much more recent. True, Dutch and North German Mennonites were involved in shipbuilding, textile production, distilling, milling, and vegetable oil production. However, their more sectarian coreligionists of South Germany, Switzerland, and France remained overwhelmingly agricultural for most of their history. Significant manufacturing again developed in the Russian Mennonite colonies in the late nineteenth century, much of it related to consumer needs and those of the predominantly agricultural economic base, in such industries as flour-milling and farm implement production.[19] However, we have no way of establishing how environmentally responsible or irresponsible these early Mennonite industries may have been.

Mennonites in North America, especially those of Swiss or South German descent or with Amish background, were slow to develop manufacturing enterprises and business enterprises generally. Their sectarian nature, fear of contamination from too much contact with outsiders, resentment of those who achieved success in business, and marked preference for agriculture all limited Mennonite entry into business and manufacturing, espccially where Amish or conservative Mennonite community control still checked individual initiative for entry into business.[20]

A few manufacturing enterprises begun as small-scale family businesses involve the processing of food products, as with the J. M. Smucker Company of Ohio (jams and jellies) and the C. H. Musselman Company of Pennsylvania (apple products).[21] While some waste products are produced in the food industry, most of these substances are natural and thus biodegradable. Although these industries, such as poultry processors, seem relatively benign, they use huge amounts of water that are disposed through normal sewage channels, ending up in the rivers, and thus become sizable polluters.

Mining is a decidedly less environmentally friendly business. It is sometimes considered a primary sector activity along with agriculture and is sometimes placed with manufacturing, as we do here. An interesting case of Amish-Mennonite involvement in mining is that of Emanuel Mullet of Ohio. Having grown up Amish on a farm, Emanuel Mullet first achieved economic success as a horse trader. He then discovered coal on his own land and began a profitable strip-mining operation there, soon expanding to buy or rent land from fellow Amishmen with the express purpose of mining the coal.

Seeing that the land was often left worthless for agriculture, his fellow Amish refused to rent or sell him more land. He was forced by the unofficial Amish boycott and unfavorable press coverage to change his methods. Fortunately, he discovered an economically feasible method of temporarily removing the fertile topsoil, strip mining the coal deposits, and finally restoring the topsoil, allowing the land to be used again for productive agriculture. Eventually, Mullet was recognized as the Coal Man of the Year in Ohio in 1986. His improved mining methods were written into Ohio mining laws, which require others to follow the topsoil replacement technique Mullet pioneered.[22] There is evidence that Mullet felt some guilt about his original mining methods, but it is questionable whether he would have developed the more ecologically sound method had the Amish boycott of his enterprise and growing negative press coverage not forced him to do so.

Notable Mennonite economic success in manufacturing has not been common, but there has been increasing involvement. One of the largest Mennonite businesses was the Hesston Corporation of Hesston, Kansas, which, at its peak in the 1970s, had sales of over $300 million per year. D. W. Friesen of Altona, Manitoba, has emerged as one of the largest Mennonite family-owned companies and is one of the largest book publishers in Canada. Sauder Woodworking Company of Archbold, Ohio, and High Steel of Lancaster, Pennsylvania, are additional examples of large Mennonite-owned corporations.

Smaller Mennonite-owned manufacturing companies include Excel Industries of Hesston, Kansas; Steiner Corporation of Orrville, Ohio; Wayne-Dalton Doors of Mount Hope, Ohio (the Mullet family mentioned above); Deweze Corporation of Harper, Kansas; Shenandoah Manufacturing Company, Harrisonburg, Virginia; Jayco Corporation, Middlebury, Indiana; Triple E, Winkler, Manitoba; Monarch Corporation, Winnipeg, Manitoba; and Neufeldt Industries, Lethbridge, Alberta.[23] There is little information regarding the environmental concerns and impacts of these industries.

Wood products and furniture making among Mennonites in both Canada and the United States have brought substantial success; such businesses include Heritage Custom Kitchens, New Holland, Pennsylvania; Loewen Millwork, Steinbach, Manitoba; Palliser Furniture, owned by the DeFehr family in Winnipeg, Manitoba; and Sauder Woodworking Company, mentioned above. Bringing entrepreneurial skill and experience with them from South Russia to Western Canada, the DeFehrs have seen Palliser grow into the largest furniture and wood products firm in Western Canada, with operations now in the United States and Mexico.

When the company added a particle board plant in the Transcona community of greater Winnipeg, fairly vociferous community opposition arose, with negative press coverage from the *East Kildonan Herald*. Community residents complained of excessive noise, smoke, acrid odors, and dust that settled on their homes and cars. The firm sponsored meetings with local residents and Manitoba provincial authorities to hear the concerns and attempt to respond. But initial efforts to solve the problem did not succeed. "We saw that no matter what we did, we were going to be criticized," DeFehr stated. "It got very ugly."[24]

Company officials began to perceive that merely meeting government standards would not be sufficient to resolve the community complaints. Many local residents had hoped to have an empty field or park where the plant was located. Frank DeFehr admits, in retrospect, that some of the complaints may have been justified. "As I look back on it, it was a positive experience. You don't need to have dust on your driveway or a squeaky conveyor running twenty-four hours a day."[25]

Even though company officials claim they were within provincial environmental guidelines from the beginning, they decided to invest heavily in new, environmentally friendly technology, including a biofiltration process that includes bacteria to consume the urea and waste wood by-products of the particle board plant, as well as eliminating most of the dust, smoke, and noise. DeFehr estimates that the company spent $1.5 million, or "about 50 percent" of the budget for the particle board plant in recent years, on biofiltration and beautification. Trees and a pond stocked with fish helped both to screen the plant from the view of passersby and to make the grounds much more attractive. "I used to get three or four [complaint] calls a day," DeFehr states. "Now I can't remember the last one I got."[26]

In this case, as with Emanuel Mullet, it may have been complaints of neighbors and negative press coverage that forced more environmentally friendly production methods. Economic factors did not operate in isolation; no company can afford to offend significant numbers of people. At some point negative press and negative public relations become costly, in both economic and noneconomic terms. To some extent, continued economic success rests on restoring positive public relations. Fortunately, that was done in both cases.

Another major furniture and wood products company with Mennonite founders is that of the Sauders in the much smaller community of Archbold, Ohio. Sauder Manufacturing Company is the largest builder of church furniture in the United States, and Sauder Woodworking Company is the largest producer and shipper of ready-to-assemble furniture in the nation. Sauder purchases particle board rather than manufacturing it, as Palliser does, but the two companies do generate a significant amount of dust, estimated by one company official at 170 to 180 tons per day.[27] Sauder has installed over 200 "bag houses" to capture most of the dust. They then burn the dust to fuel an electric cogenerating plant and are able to sell electricity to their local electric utility. Steam turbines produce the electricity, and some steam also provides heat for the plant. None of the dust or scrap is taken to landfills, although there are some problems with disposal of waste paint thinners.

Although the company did invest thousands of dollars in bag houses to capture dust before they were required to do so by federal and state regulations, management does tend to see Environmental Protection Agency regulations and monitoring as burdensome. "We had invested a quarter of a million dollars each in thirty-three bag houses before we even heard of the EPA," says John Schlatter, company spokesman. "They are constantly watching over our shoulder. Regulations are tight. The EPA has multiple regulations, and they apply per facility, not per business."

The company also spends major sums on air sampling, called *stack testing*. They hire private firms to do stack testing to check for particulate emissions, mostly of wood dust, to make sure they are within allowable standards. Sometimes EPA officials must witness the tests that Sauder is paying for. "It's an ongoing challenge to make them happy," Schlatter says, "Sometimes they ask for a higher level of compliance than they have a right to."[28]

Unlike Palliser's public relations problems with area residents, however, Sauder has had few complaints from their local neighbors, perhaps partly because they and the Archbold community have grown together and many local residents are employed by Sauder. "We get no complaints from Archbold residents about noise, odor, or dust."[29] John Schlatter says that the only recent local complaints have come from landowners who oppose Sauder's efforts to get a cloverleaf interchange on the Ohio Turnpike close to the plant.

From the limited data available, it seems that Sauder is making good-faith efforts to be environmentally responsible, but it is also evident that they are well aware of the costs of doing so and regard EPA regulations and monitoring as somewhat burdensome. In this regard they are probably not unlike non-Mennonite businesses. Sauder has grown and is no longer an entirely Mennonite firm. Marvin Lantz estimates that, of their nearly three thousand employees, only about a quarter are Mennonite. Mr. Schlatter is one of the increasing number of Sauder employees who are not Mennonite.

Several firms have been founded by environmentally conscious Mennonites to produce and market modern environmentally responsible products. One of the first of these was Sunflower Energy Works, founded in Lehigh, Kansas, in 1977. It produced solar panels, some of which were installed to heat buildings at Hesston College. The company was profitable as long as a federal subsidy in the form of a deduction from the income tax of purchasing individuals and businesses encouraged such purchases of its products. However, when the federal income tax deductions were ended, sales dropped, the business was no longer profitable, and it was closed.[30]

Similar Mennonite-owned companies were formed to produce solar products in Ontario in 1982 and in Harrisonburg, Virginia, during the same period. Both flourished when respective Canadian and U.S. tax credits were available, and both also became unprofitable and were closed when the credits were terminated.[31] A more recent attempt to develop and market environmentally friendly products is being made by a small company called Jetstream Power International in Ohio. It is developing photovoltaic cells, solar-powered water pumps, and windturbine generators. The targeted markets are mostly in remote areas of North America and overseas, where conventional power sources are not available, but clearly there is an attempt here to harness wind and solar energy in an environmentally responsible way and at the same time to make a profit. It is significant that the founder and owner of the firm, Clarence Stutzman, is a former Mennonite Central Committee volunteer with overseas experience who sees himself as on a now-private mission to make environmentally friendly technology available to poor persons in North America and abroad. It is too early to tell whether this business will become profitable and enjoy long-term stability and growth.[32]

THE THIRD ECONOMIC SECTOR:
SERVICE INDUSTRIES

A majority of modern workers, including Mennonites, are employed in various service industries. There are hundreds, if not thousands, of such industries and certainly thousands of job titles. With minimal Mennonite involvement in manufacturing and decreasing numbers of Mennonites employed in agriculture, they may have become even more dependent on service industries than the population at large in recent years. Especially notable is the extremely high percentage of Mennonites in professional and technical job categories. By 1982, a census of the Mennonite Church found that adult, employed Mennonite women were more than twice as likely to be active in professional and technical categories as was the population of American women as a whole, and Mennonite men were also significantly more so.[33]

Using slightly different categories and a sample of five major progressive Mennonite groups in the 1989 church member profile, Kauffman and Driedger found that 36 percent of Mennonite men and 21 percent of Mennonite women were in professional occupations and that another 16 percent of men and 4 percent of women were in managerial positions, probably mostly in service industries. Partly because of a strong service ethic, many Mennonites have done terms of voluntary service both at home and abroad, and many have entered such professions as teaching, medicine, social work, and Christian ministry. Increasing numbers are employed as accountants and salespersons, and a few are even lawyers. There are, of course, many Mennonite transportation workers, factory workers, food service workers, and clerks.

Many Mennonites are employed in the construction industry as both workers and owners/managers. An example is High Industries of Lancaster, Pennsylvania, a large steel construction and real estate company, and there are others not as large. It would be very interesting to determine whether Mennonite construction and architectural firms urge clients to use energy-conserving materials and environmentally friendly designs in the construction of homes and businesses. One Mennonite design firm in Saint Joseph County, Indiana, owned by LeRoy Troyer, does attempt to design environmentally friendly housing projects. However, the available data are not adequate to draw conclusions about the extent of environmental concern in the Mennonite-owned construction industry.

Do Mennonite farm implement dealers urge farmers to adopt environmentally friendly implements and tillage practices and rely less on energy and chemical-intensive farming? Do Mennonite-owned and -managed hospitals and medical clinics dispose of medical waste any differently than non-Mennonites? Do Mennonite food processors, such as Moyer Packing Company of Souderton, Pennsylvania (one of the largest meat packers in the United States), use procedures and treatments that are any less polluting than those of non-Mennonite peers? These are interesting questions, but data are lacking.

Within the large Mennonite nonprofit services sector, the Mennonite Central Committee stands out as especially noteworthy. While it has an educational mission of sorts, its primary mission is relief and development, mostly in less-developed countries and in less-developed areas of North America. The Mennonite Central Committee has promoted appropriate technology in most of its overseas development projects, such as irrigation pumps using a bicycle frame and human power rather than a fossil fuel–powered pump that would cost much more both initially and for continued operation. Other examples include hand-cranked peanut shellers and the manufacturing of building blocks using locally available clay and human power rather than purchasing blocks built and sold by a commercial block factory. In many cases the chief objective is to give poor people access to technology they can use and afford. In almost all cases, however, this technology is environmentally friendly. Because it is "low-tech" as opposed to "high-tech," chemical and fossil fuel dependency is minimized, along with cost.

Although the greatest influence of the Mennonite Central Committee (MCC) is overseas, it definitely has an effect in North America as well, both in projects conducted directly in Canada and the United States and indirectly as volunteers return to home congregations and home communities. One example is the phenomenally popular cookbook by Doris Janzen Longacre, *The More with Less Cookbook*, and its sequels, *Living More with Less* and *Extending the Table*.[34] Yet another example would be the growing number of self-help stores that sell to North American consumers various hand-crafted items produced by poor artisans abroad, with the lion's share of the proceeds going back to the producers because both the store clerks in North America and the MCC personnel overseas are volunteers. In some cases, these stores also function to recycle clothing and other consumer items, certainly an environmentally friendly function as well as one that furthers social justice. These self-help stores have moved beyond Mennonite circles.[35]

Through its educational promotion of environmentalism, MCC has nurtured numerous conservation and recycling actions. It pioneered the first "blue box" collection and recycling program in Kitchener, Ontario, in the mid-1970s, which has now spread to many cities across Canada and the United States. Under the leadership of Dave Hubert, MCC developed the Edmonton Recycling Society, which has been collecting, processing, and marketing garbage for half of Edmonton since 1988. The Edmonton Recycling Society has refunded the city of Edmonton as much as a half-million dollars a year from the profits of this "nonprofit" operation. The "economic impact of the recycling operation on western Canada is now about $8 to $11 million annually."[36]

In eastern Canada, "a corporation supported by the Mennonite Central Committee of Canada [was] chosen to manage the recycling of scrap tires" in 1996, to be established in New Brunswick and Nova Scotia. Approximately 1.4 million tires are to be recycled each year. The MCC-sponsored company was the only church-related bidder, and it got the contract.[37]

Clearly, the MCC could do more. With broad and deep support from virtually all Mennonite groups demonstrated by cash contributions, donated material goods, and countless hours of volunteer labor, MCC has real economic advantages enjoyed by few other nonprofit agencies, let alone for-profit firms. On the other hand, the MCC is a reflection of its constituency and cannot be too far ahead of its supporting members for fear of alienating them.

What are Mennonite institutions of higher learning doing to be environmentally responsible, and how does economics affect their efforts? In the absence of an exhaustive, systematic survey, I used a fairly new technology called electronic mail, which is cheaper and faster than surface mail, to survey colleagues at Mennonite institutions.[38] First, all of the Mennonite colleges from which significant information was received have some type of course work available to students in the environmental studies/environmental sciences area. As of December 1999, Messiah has both a major and a minor in environmental science. Eastern Mennonite University has a major in environmental science in the biology department, a course in environmental and conflict transformation in their new master's degree program in conflict analysis, plus a major in international agriculture. Goshen College has a major and a minor in environmental studies. Bethel College has had a major in environmental studies since 1970. Fresno Pacific University offers majors and minors in environmental science and environmental studies. Tabor College offers a major in environmental biology.

Most Mennonite colleges now have campuswide recycling programs in operation, and some (such as Goshen) provide drop-off sites for the general community in which they are located. Several caveats are in order here, however. In many cases recycling is now mandated by state or local law. Campus and community recycling is fast becoming the norm rather than the exception, in non-Mennonite as well as Mennonite communities and campuses.[39] The public schools are also helping to spread the recycling gospel.[40]

Furthermore, spokespersons for several colleges indicated that the impetus for campus recycling had come from student initiative and that the program is operated wholly or almost entirely with volunteer student labor and thus costs the colleges very little. Typical is a response from Bethel: "Our voluntary recycling program was started five years ago by students. This program, though, is done at a small cost to the college, perhaps $1,000 a year for work-study help."[41] Similarly, several persons who are or have been at Goshen reported that the recycling program there was begun on a small scale by students and that it took several years to convince the administration to make the program official, despite very low start-up and operating costs.[42]

Many colleges have undertaken the retrofitting of buildings, often including insulation and window replacement, to save energy costs. Here economic considerations clearly mandate action, especially with older, energy-inefficient buildings. Wheaton College economics professor (and Mennonite) Norman Ewert advised that investment in insulation and energy-efficient lighting would be recovered in one year, and Wheaton withdrew endowment funds to cover the costs. As predicted, the costs were recovered within twelve months.[43] In most cases the pay-off period is longer, but most colleges come up with the funds rather quickly when they can realize savings from increased energy efficiency in just a few years.

Administrators are understandably less willing to commit scarce funds to projects that do not have a clearly anticipated and direct payoff, especially when budgets are tightened in times of reduced enrollment and retrenchment. From the perspective of Goshen biology professor

Stanley Grove, "we are caught in a mode that looks primarily on the bottom line and on a short-term scale. Any changes are resisted unless they will save or make money."[44] Goshen College was reluctant to accept a gift of land intended by the donor to become an environmental preserve and educational center until the college received sufficient endowment to guarantee that operating what is now the Merry Lea Environmental Learning Center would not be a drain on the annual operating budget or on general endowment income.

The center now has an annual operating budget of $400,000, which is kept separate from that of the college itself. The separate endowment of $3.3 million provides the operating revenue needed each year. The center provides an environmental learning and research laboratory not only for Goshen students and faculty, but also for the larger Northern Indiana community. Yet, says center Director Larry Yoder, "Reservations [still] come in the form of whether or not Merry Lea helps GC fulfill its mission . . . My personal belief is that the environment will be the central paradigm for the 21st century—politically, economically, and theologically . . . In my opinion, GC has not yet made the shift to green in its gut." Yet, he adds, "GC is ahead of the curve."[45]

Perhaps the most ambitious current project in the environmental area on a Mennonite campus is the construction of a cogenerator at Fresno Pacific, as reported by administrator Bruce Traub:

We are initially installing three 120-watt generators driven by John Deere engines running on natural gas. We will add engines/generators as the campus expands. The system is both a central [heating] plant and a co-generation operation. While I was interested in reducing electricity costs, the bigger interest and more significant impact is to reduce overall utilities costs, and to do this requires that the entire system be engineered as a heat producer, not a generator of electricity. Hot water either heats the building directly as well as providing heat for domestic hot water, or we run the hot water through an absorption chiller to produce cold water to cool the buildings. We will have a four-pipe hydronic network to supply both hot and cold water at all times to each connected building.[46]

It is somewhat ironic that the colleges within the Mennonite orbit that seem to have made the most concerted recent efforts to reduce their environmental impact are Fresno Pacific and Messiah, the least Mennonite, in terms of their student bodies, of the colleges polled. In fact, neither of the persons reporting from these two schools, Bruce Traub at Fresno Pacific and Joe Sheldon at Messiah, is Mennonite or claims Mennonite background.

Clearly, environmental sensitivity and commitment are not limited to Mennonites. The recently released *Redeeming Creation: The Biblical Basis for Environmental Stewardship* has nary a Mennonite author.[47] As Mennonites and Mennonite colleges become less sectarian, they are becoming more willing to cooperate in the environmental and other areas of concern with fellow Christians of various denominations. For example, most four-year Mennonite institutions of higher learning now belong to the Coalition of Christian Colleges and Universities (CCCU), for which Myron Augsburger served as president until recently. This somewhat ecumenical but generally evangelical collection of Christian colleges and small universities sponsors cooperative learning opportunities for students and faculty of member institutions. One of these is the Au Sable Institute, an environmental center at Mancelona, Michigan. Of the forty participating colleges, mostly Midwestern and members of the CCCU, Goshen, Tabor, and Messiah are the three Mennonite-related institutions that directly participate in funding and operating Au Sable. Faculty and students of other CCCU institutions, including Bethel, Bluffton, Eastern Mennonite, and Fresno Pacific, are also eligible to participate through their membership in the CCCU.[48]

The participation of Mennonite institutions in the CCCU generally and the Au Sable Institute specifically is a sign that Mennonites no longer wish to "go it alone" in the environmental area. With regard to the Mennonite witness to pacifism, we may still be somewhat unique. But environmental consciousness and commitment are not limited to Mennonites. And economically, programs such as that of the Au Sable Institute are costly. When costs can be shared, by institutions both Mennonite and not, costs per sponsoring institution are lower.

CONCLUSION

There are no systematic hard data to determine whether Mennonites are more environmentally conscientious than others. Mennonites, especially those living in middle-class North America, make compromises for creature comfort all the time. Those of us with serious commitments to environmentalism in our daily lives will often feel that we are swimming upstream, and we need to realize that such commitments will often be costly, as the Schrock-Shenks have discussed:

> Carolyn and I recently rehabbed a row house in Lancaster. We put a lot of money into Loewen windows from Canada. We got triple-pane, double low-e windows with a whole unit R factor of 6.3. They were very expensive. Our contractor thinks we're nuts. Sometimes I wonder myself. We studded out the exterior walls and were planning to use foam insulation, which would have given us an R value of 21+, which would have exceeded the government's recommendation for new construction. We ended up using cellulose because of the cost of foam, giving us an R value of 11+, which is the recommendation for remodeling. We are now going to buy a used refrigerator instead of a new, CFC-free model because we ran out of money![49]

This couple shows an admirable commitment to environmental stewardship. But even they have been forced to compromise their ideals for economic reasons. And how many of us would have this level of commitment? My guess is that most Mennonites would have compromised a good deal earlier than the Schrock-Shenks.

The Reformation generally and the Anabaptist movement especially were attempts to restore the true Church of Christ, to purify what had become a sadly corrupt and impure church. Compared to Lutherans and Calvinists, who are more likely to be concerned with doctrinal orthodoxy (i.e., correct Christian belief), Mennonites have put relatively more emphasis on orthopraxy (i.e., right living).[50] Even our church splits have often been over behavioral and lifestyle issues rather than doctrinal orthodoxy per se, further evidence of how seriously we take lifestyle issues.

It is natural, then, that Mennonites should have real concern for right living with regard to responsible care of God's creation, along with other issues of lifestyle and stewardship. Although there may be cases where the Mennonite environmental track record is good or at least better than average, as with the relief and development efforts of the Mennonite Central Committee, my overall conclusion is that we are rarely able to ignore the economic "bottom line" referred to in my father's poem. Whether as farmers, manufacturers, or service providers or in our daily lives, the bottom line often causes us to behave in ways not significantly different from choices made by our non-Mennonite and even non-Christian neighbors.

Most of us have chosen to join the North American social and economic mainstream. We like the comfortable North American lifestyle, even if it is extremely dependent on high energy consumption and use of nonrenewable resources. We live in a "high-tech" society. Some of that technology, such as minimum tillage in agriculture and windturbines and solar panels as power sources, may help us preserve the environment, but much of it (e.g., the family car, air conditioning, disposable diapers) is environmentally unfriendly.

As our opening illustration from Paraguay should teach us, the Western model of development to which most of us are wed continues to take a terrible toll on the environment. As we have "developed" our farms, our factories, our service industries, and our personal lives and careers, we have, in fact, taken an increasingly higher toll on God's creation. Perhaps it is time to reconsider the "Amish option." Perhaps it is time to consider applying MCC's low-tech "appropriate technology" in North America, not just in Africa and Asia. How many of us are really willing to accept the cost?

CHAPTER 6

The Mennonite Political Witness to the Care of Creation

Mel Schmidt

"If we Mennonites keep on being *die Stillen* much longer, there won't be any *im Lande* left." This was the oft-repeated battle cry of Jo Androes, a Mennonite political activist in Wichita, Kansas, from the early 1970s through the mid-1980s. Androes had major difficulties with the classic self-definition of Mennonites as *die Stillen im Lande* (the quiet in the land). She practiced what she preached, managing several successful campaigns for Governor John Carlin in Sedgwick County and going on to be the state coordinator for Tom Docking in his 1986 campaign for governor.

Jo Androes's involvement in politics was triggered by her concern for the environment. A longtime Republican, she changed parties in the early 1970s because she realized that her support of Republican candidates was almost always geared toward finding the candidate who would do "the least damage" environmentally and socially. She believed that the Democratic party was more open to positive governmental action to save the environment. Androes had earlier become a member of the Greenpeace movement because of its efforts to save the whales. Her Greenpeace involvement caused a growth in her awareness of environmental issues, which in turn triggered a change in her political affiliation.[1]

Any examination of Mennonite political involvement in environmental concerns must begin with several caveats. First, we must recognize that a person's motives for political involvement are usually complex and not necessarily driven by single-minded concentration on an issue. Jo Androes was politically active for about fifteen years for many reasons other than environmental concerns. In fact, having known her personally during some of those years and been involved with her on several political projects as her pastor, I must confess that the environmental question never came up. I would not have suspected that she was deeply concerned with environmental issues, which had not yet appeared on the public radar screen. I did not know then, as I know now, that she was an organic gardener with deep concerns about chemical pollution of the food chain and a member of Greenpeace. As her pastor, I knew her to be a political activist who was continually challenging her congregation to be involved in the political process on various issues. Her challenges eventually bore fruit. A committee was formed to study issues and recommend action. One such action in the early eighties was the publishing of a slate of endorsements for various political offices by the Lorraine Avenue Mennonite Church in Wichita, Kansas.[2]

A second caveat is that political action takes place in many different arenas and is not necessarily related to legislation. Political involvement to save the environment may be as local as working with the district school board, where a citizen might be deeply involved in promoting environmental education among schoolchildren. A Mennonite farmer might become active politically to save his farmland from being turned into a landfill or a missile site. The many ways in which environmental political activities may be interrelated with each other have been discussed in *Hope for the Family Farm.*[3]

Keeping these caveats in mind, any investigation into the question of where and how people of Anabaptist persuasion have been involved politically in environmental issues will lead to an examination of a variety of contexts and situations. One should expect to find vast differences in the level of political action depending upon many factors, including the immediacy and urgency of the environmental problems being addressed. For example, Mennonite farmers have belonged to rural cooperatives for many years and have also joined such groups as the Farm Bureau and Soil Conservation Service Districts, all of which are involved in political matters on the national level. Some political involvements are environmentally motivated in the sense that preservation of farmland, the family farm, and rural life per se all imply preservation of resources to make such a life possible.[4]

CHURCH–BASED INVOLVEMENTS IN ENVIRONMENTAL POLITICS

A description of Mennonite church-based involvement in environmental politics requires that one first define terms. For example, it can be argued that Mennonite resistance to and refusal to participate in war is a historical commitment to saving the earth, simply because war is the greatest polluter known.[5] While it cannot be denied that war is the greatest polluter on Earth, I choose to define the practice of environmental politics more narrowly. We will assume the necessity for conscious focus on the saving of the earth, recognizing as we do that the goals of peace and justice for humankind are interwoven with the saving of the earth. If the earth is allowed to go on its present course toward destruction, there will indeed be very little peace and justice left for anyone, as Jo Androes so often warned.

The Fort Riley Expansion

Our first example of corporate Mennonite environmental political action is, indeed, ambiguous. In early 1989 the Fort Riley Installation in Kansas made public a proposal to purchase 100,000 additional acres of farmland to extend its existing 100,000-acre facility. Living on the land considered for the expansion were 320 families who would be displaced by the expansion. The military authorities argued that the purchase was necessary because current practice fields were filled with undetonated artillery. Rather than cleaning up the facilities already owned, the government wanted to expand into new areas. This was decried by opponents as irresponsible.

The issue quickly became multifaceted, involving issues of the environment, peace, justice, and rural life. On April 6, 1989, more than 750 local landowners and farmers formed the Preserve the Flint Hills Citizens Coalition to protest the expansion of the military base. The Interfaith Rural Life Committee of the Kansas Ecumenical Ministries voted unanimously in its April meeting to support the farmers. The Mennonite Central Committee was represented in this group and thus became part of the supporting cast for the farmers who were trying to save their land.

The resolution presented to the delegates gathered at the 1989 annual meeting of the South Central Conference of the Mennonite Church supported the landowners, but the rationale for the support was entirely based on biblical perspectives regarding the stewardship of the land. The "be it resolved" part of the resolution asked for only three things, all focused on environmental issues: (1) Fort Riley must be required to clean up its practice fields, (2) the purchase of additional land must be blocked, and (3) the Environmental Protection Agency must be asked to assess the environmental impact on all land currently in the control of Fort Riley.[6] The fall of the Berlin Wall in November 1989 and the subsequent end of the cold war resulted in the downsizing of the military procurement program in the United States, so the expansion plans at Fort Riley never materialized.

The Preservation of Mill Creek Valley

Beautiful, peaceful, bucolic Mill Creek Valley is located just south of New Holland, Pennsylvania. It is filled with traditional Amish farms. The land itself is probably as fertile as any land in the entire continent of North America and needs no irrigation because of an exceptionally high water table. In 1993 the Eugene Eberly farm was to be sold to a group of investors for the purpose of creating a large retirement center. Mr. Eberly was ready to sell. He wanted to cash in his farm and retire. Mrs. Eberly had cancer and her medical bills were mounting. The Eberlys assumed, as most Americans do, that they could sell their property whenever they wanted and for whatever price they could get. The Eberlys had faithfully taken care of the land for many years and had watched their fixed assets grow to the point where they could sell the land, retire on the proceeds, and pay their substantial medical bills.

Plans for the Garden Spot Village Retirement Center were circulated in the community via flyers and the local newspaper. The entire Eberly farm, to be purchased for $2 million, would eventually be developed into a large retirement community, one of many such communities dotting the landscape of Lancaster County.[7] The land south of New Holland was zoned for agricultural purposes, so rezoning would be necessary. The three township officers, who supported the development of the area, scheduled the first public hearing on the matter, anticipating no problems. The meeting was to be held in Martindale, about eight miles north of Mill Creek Valley.

The township trustees had not counted on one thing, the reaction and subsequent involvement of Amish farm owners and their neighbors. Edna Esh, a diminutive, single Amish woman who owns a small tract of inherited land in Mill Creek Valley, learned about the rezoning meeting in Martindale only two days before the first public hearing was scheduled.[8] She also learned something else that sent her blood pressure skyrocketing. A sewage plant, which was necessary for the planned development of the retirement center, was to be built on her land.

It was clear to Edna what she must do and that it must be done at once. She set out on foot to visit her neighbors, all eighty of them, telling them about the plans for the sewage plant and retirement center. She drew up a petition against construction of the sewage plant and had people sign it. She encouraged her neighbors to attend the zoning meeting, even though very few of them had ever attended such meetings and very few had bothered to inform themselves of what was about to happen in their beloved valley. The plain people had to hire drivers to go to Martindale for the evening meeting because it was too far for them to travel by horse and buggy after dark. Nearly five hundred people came to the hearing, which resulted in the tabling of the motion to grant a license for the sewage plant. Subsequent hearings held throughout the fall of 1992 finally resulted in a reversion of the zoning back to its original status.

That meant that the Eberly farm could not be sold to the Garden Spot Retirement Center investors. They subsequently built a scaled down version of the originally planned facility near the northern edge of the valley, where no new sewage plant was needed. A significant result of the attempted assault on Mill Creek Valley was the formation of the Earl Township Farmland Preservation Trust, which is still working to preserve Amish farms in the New Holland community. The trust is an important part of the local political scene. Members meet regularly and keep themselves informed on development efforts, and they push for zoning laws so that the Amish way of life can be protected from the encroachments of residential development.

Three farms in the valley have been awarded the "preservation" designation, which means that these farms cannot be sold for residential subdivision. Some landowners, Amish and others, are reluctant to obtain the designation because it locks in a lower value for their property. Several Old Order Mennonites have already moved out of the community to Wisconsin. Such families are interested in maximizing the sale of their property and thus are not necessarily interested in the preservation of farms.[9]

The Kickapoo Valley Low-Flight Zone

In February 1995, the Wisconsin Air National Guard announced a plan to expand its low-level training flights in southern Minnesota and northern Iowa. The expansion plans included an enlargement of the Hardwood bombing range on the border of Wood and Juneau Counties near Nekoose, increasing the range by 7,137 acres. An existing training route beginning near Rochester, Minnesota, would be kept, and sorties would be increased from 2,187 to 2,423 per year. Two new training routes beginning in northern Iowa would be established, with a planned 2,151 sorties per year on the new routes.

By October 8 the *Wisconsin State Journal* reported that opposition to the plans of the military had spread to at least four counties and eight municipalities, plus other groups such as pilots associations, farm groups, environmental groups, and political organizations.[10] In addition, the Wisconsin Conference of Churches (WCC) sponsored a Rally for Peaceful Skies at the Capitol in Madison on October 30, 1995. Aaron Rittenhouse, a Bethel College graduate, spoke as a representative of the Mennonites. He had been invited by the WCC because of a desire to have Mennonites involved in the movement, even though there are few Mennonites living in Wisconsin and they are not formally connected to the WCC.

For the hearings, the opposition brought in Rear Admiral Eugene Carroll, director of the Washington-based Center for Defense Information. Carroll's arguments took on the military on its own turf by maintaining that low-level flight training is obsolete for the kind of wars fought these days. "The issue has been clear since Vietnam," said Carroll. "The need for low-level training is, I think, contrived." Thus, the public hearing process ruled out the expanded training routes, although

Dennis Boyer, a longtime environmental activist in Wisconsin, maintains that the military probably "got what they wanted."[11]

Once the military had given up its plans for low-level training flights, the massive groundswell opposed to the rest of the plans for the expansion of the bombing range evaporated. Boyer believes that the military planners floated the plan with the intention of giving up the flight corridors as a bargaining chip to keep what they really wanted, which was an increased bombing range. In that sense the celebration over the military's "defeat" regarding the flight corridors is premature.

Probably about a dozen members of the Madison Mennonite Church (Madison, Wisc.), became involved in the movement in various capacities. Events were regularly announced to the congregation, and concerns were shared during prayer time. Within the church's membership are professional people like Aaron Rittenhouse, a history teacher, and Dennis Boyer, an attorney with years of political experience as a labor lawyer. Boyer has significant contacts in the political arena across the state and also has some personal dealings with the Amish community because he breeds Percherons and visits Amish horse breeders regularly. Through Boyer's contacts the Amish became involved in the movement to ban low-level flying. They broke their tradition of avoiding public controversies by sending a fifteen-page letter, "in perfect hand writing," outlining their concerns about startled horses and the general nuisance caused by low-level flights.[12]

Some Generalizations

The main "generalization" to be drawn from these examples is that no generalizations are possible. For example, it is tempting to assert that Mennonites, like most people, are victims of the NIMBY (Not In My Back Yard) syndrome. They will get involved in political activity only when their immediate interests are threatened. This would seem to be the lesson learned from the Mill Creek Valley action, in which Edna Esh, who had never voted and had never expressed any political opinion publicly, nevertheless became a very effective political organizer. But the NIMBY generalization will not hold up for either of the other two case studies. Neither the Mennonites in Kansas nor the Mennonites in Madison, Wisconsin, had overwhelming personal reasons to become involved. There were no Kansas Mennonite farmers threatened with the loss of land. No low-level flights were ever planned anywhere near Madison, Wisconsin. In the Wisconsin movement, however, the Madison people brought the Amish into play through a process of gentle, personal persuasions. The Amish bishops had obvious personal reasons to be involved in the project. The Madison congregation became a liaison between the wider Christian community and the isolationist Amish.

Another tempting generalization is that Mennonites will usually not take the lead in public controversies but will gladly join coalitions in causes that are compatible with Mennonite beliefs. This would seem to be the lesson in both the Fort Riley and the Kickapoo Valley movements. In both cases, it seems that Mennonites "signed on" to a political action when it had already matured into a full-blown movement with good chances for success. The initiatives for both movements began somewhere else. However, the Mill Creek Valley situation teaches the opposite, which is that Mennonite people will at times take up a minority position, regardless of how hopeless it seems at the beginning.

INDIVIDUAL INVOLVEMENTS

Any description of Mennonite participation in environmental politics would be incomplete without at least an attempt to render a sampling of individuals who are politically involved in significant ways—locally, provincially, nationally, and in movements.

A Politician for Recycling

Everett Thomas is president of the Mennonite Board of Congregational Ministries of Elkhart, Indiana.[13] He is also a local civic leader, serving his second term on Elkhart's city council. During his first run for office, he made environmental issues the centerpiece of his campaign. He proposed curbside recycling in 1991, when recycling was considered radical in political terms. He also pushed hard for bike paths so that people would be encouraged to use bicycles instead of cars. During Thomas's first term in office, curbside recycling was put on the agenda at his initiative. His success in being elected on that issue had made it discussable for the first time. Curbside recycling, however, was an issue whose time had not yet come for Elkhart. Drop sites for recyclable materials were tripled in an effort to make recycling more convenient without making it mandatory for everyone.

During his campaign for reelection, Thomas's opponent ran against him on one issue—doing away with alternative transportation, an issue in which Thomas had heavily invested. His opponent, a Republican in a heavily Republican city, apparently felt that Everett Thomas, a Democrat, was vulnerable on that one issue. However, Thomas won a second term and went on to get involved in other environmental issues, such as water quality and land use. Walmart sought to build a superstore on eighteen acres of prime farmland at the edge of the city. Brought to a vote in the city council in October, shortly before the election, the Walmart was voted down 4–3. This gave the opposition an election issue. Council members who had voted against Walmart were painted in broad brushstrokes as being "anti-development." Two "pro-development" members were elected, and the new council voted again on the Walmart proposal. This time the proposal passed 6–1, with Everett Thomas casting the lone dissenting vote.

An Administrator for the Environment

David Neufeld is senior policy and program planner for the Ontario Ministry of Environment and Energy. In over a decade of governmental employment since 1985, he has worked for three different political parties. The present governing party has already cut 50 percent of the budget for environmental protection. One-third of the staff in the Ministry of Environment and Energy has been cut, and more cuts are promised. While recognizing the discouraging times in which all environmental political activists work, Neufeld gamely applies himself to the task, considering himself fortunate to have found a way to work in his chosen field. "Through it all," Neufeld says, "I have resisted the temptation to rationalize the compromises one faces in my position, to ignore the contradictions, or to succumb to the constant pressure to give in to opposing interests. At the same time I recognize that one needs a long-term vision and faith to sustain one when hard-won gains seem to suddenly slip away."[14]

Neufeld grew up on the prairies of Saskatchewan and spent much of his childhood enjoying the world of nature. He fantasized about becoming a rancher and herding cattle across the prairies. Then he became interested in replanting the forests of Canada, and from there his interest became park planning. He received a bachelor's degree in environmental studies in 1983 and finished a master's degree in regional planning and resource development, both from the University of Waterloo. In 1998 he completed work on his Ph.D. in Environmental Planning at the University of Waterloo. His work has been published in journals, and he has read papers at conferences on subjects such as environmental planning, land use, ground-water protection, and hazardous waste management.

A Registered Lobbyist

An early encounter with a native black-footed ferret, now apparently extinct in the United States, confronted David Ortman with a basic environmental question: Why should the black-footed ferret be preserved, when it has no apparent economic value?[15] Ortman decided that it was just as important to keep the ferret as it is to keep cows, of which there are millions. Because of their obvious economic benefit, cows are in no danger of becoming extinct. Ortman realized early that some animals and plants have to be saved from the destroyers even if economic value cannot be proven.

Armed with a degree in environmental studies from Bethel College, North Newton, Kansas, David set out for Seattle to begin a voluntary service position with the Northwest Office of Friends of the Earth (FOE) for 1975. A love affair with his work became his professional commitment. He is now director of the Northwest Office of FOE and a registered lobbyist with the U.S. government. Ortman likes to say that "politics is a lot like milking a cow. Much of the work is done sitting down, there are always more 'handles' than you can get hold of, often a peculiar odor permeates the air, and finally, the finished product is amazingly elusive and hard to grasp."[16]

Despite his modest self-appraisal, David Ortman has built an impressive track record in efforts that are tangible and not so hard to grasp. He worked on the Alaska Lands bill that resulted in 100 million acres of federal land in Alaska being set aside for national parks, monuments, and wildlife refuges. In 1977 he pioneered a resolution on Christian Stewardship of Energy Resources, which was adopted by the General Conference Mennonite Church in Bluffton, Ohio. This seems to have been the first such statement by any church denomination. In 1988 he drafted legislation that resulted in the establishment of Grays Harbor National Wildlife Refuge in Washington. This effort involved much media work, lobbying in Washington, D.C., and testifying before congressional committees. In 1993, when the Asia-Pacific Economic Cooperation group met in Seattle, Ortman helped put together a coalition of environmental and labor groups for a press conference on the hidden costs of free trade, which was attended by three thousand media people from around the world.

Ortman recognizes that there is little in his background that would lead one into political activity. "If someone would have told me while I was attending a one-room rural schoolhouse for eight grades that I would become a registered lobbyist for a national environmental organization, testify before Congress and meet with Senators and Representatives, it would have sounded like science fiction." He continues to challenge the body of Christ as well as the body politic on environmental issues. He believes the church is already lobbying when language about "a new heaven and a new earth" is freely used. "What good is a church," he asks, "without a habitable planet to put it on?"[17]

A Science Coordinator for Oil Spill Cleanup

"It's for the birds," says Stan Senner, when asked about why he became an environmentalist. He is an avid birder, having been the director of the Migratory Bird Conservation Program for the National Audubon Society.[18] He is now science coordinator for the *Exxon Valdez* Oil Spill Trustee Council, which is responsible for spending the $1 billion cleanup fund established by the Exxon Oil Company as part of the environmental damage settlement.

After his graduation from Bethel College, North Newton, Kansas, in 1973, Senner spent one year in Mennonite Voluntary Service in Seattle working for Friends of the Earth. He was replaced by David Ortman in 1974. Stan moved to the Fairbanks Environmental Center in Alaska and took up graduate studies at the University of Alaska, after which he went to work for the Wilderness Society in Alaska, lobbying Congress for the Alaska Lands bill to establish 100 million acres for parks and refuges. In 1979, Stan went to work for Congressman Tom Evans of Delaware as a staff member for the Merchant Marine and Fisheries committee. He worked on fisheries and wildlife legislation for three years (1979–82) and then became executive director of Hawk Mountain, a private wildlife sanctuary in Pennsylvania.

After the *Exxon Valdez* oil spill, Stan visited Prince William Sound and researched the damage done to birds. Governor Cowper appointed him to work on oil spill damage control. When Wally Hinckel was elected governor of Alaska, Stan's job was over. He moved to Colorado and worked for the Audubon Society and then, in March 1995, he was called back to Alaska by Governor Tony Knowles to work on oil spill cleanup.[19]

An Environmental Lawyer

Carolyn Raffensperger calls herself an environmental lawyer and seeks always to keep one foot in the church and the other in politics. She has facilitated conferences and forums on the environment. For the Service Employees International Union meeting in Washington, D.C., on April 16, 1996, she brought in Herman Daly to be the speaker. Carolyn joined the Mennonite Church in Chicago as an adult and is a member of the Joint Environmental Task Force. She is an archaeologist, attorney, and self-described "congregational environmental activist." She has been involved in helping to organize a group to protect Amish farmland in eastern Ohio, where the "Lancaster County disease" is spreading and threatening to decimate Amish communities. Raffensperger lives with her husband, Fred Kirschenmann, on a farm in North Dakota.[20]

AS CAN BE SEEN FROM THIS BRIEF SAMPLING of individual environmental activists, "many roads lead to Rome." These individuals are very clear about the fact that there is no single way to protect the environment. There is no single arena in which the game must be played. All of the people described have great gifts of versatility and flexibility. They are adept at responding to a bewildering array of issues. They exhibit an extraordinary ability to communicate with common citizens as well as government officials and academics.

No serious environmental activist is blind to the need for legislation, but neither will he or she discount the need for basic education in local settings such as the church, school, or social club. The field is literally the world, and the world itself is the arena for the work of environmental action. Nothing can be outside the concern of an environmentalist, whether it is the smallest creature (black-footed ferret) or the largest seacoast (Prince William Sound).

A CRITIQUE

By their faithful track record on peace and justice issues, as well as their historical love of the land, Anabaptist/Mennonite faith communities have earned the right to speak out on environmental issues but are quite content to be *die Stillen im Lande*—an irony of our time. This is puzzling and mystifying, particularly in view of the fact that even the most isolationist groups among them will dig in their heels and take tough political stands on controversial issues when the need is clearly present.[21] We have seen both the reticence and the willingness in our case studies. In Wisconsin it became very clear that Mennonites were coveted as a presence in the movement, even though their numbers in the state are pitifully small.[22] According to Dennis Boyer, the well-publicized activities of Mennonite Disaster Service after natural disasters gave the Mennonites a certain credibility, visibility, and integrity unique to them.[23]

Even more puzzling, perhaps, is the nearly total absence of any identifiable Anabaptist/Mennonite political activity in an area that one would think is near and dear to their hearts—sustainable agriculture. For example, there is no institutional Mennonite presence in the Sustainable Agriculture Coalition, a Washington-based, national umbrella organization that lobbies for legislation and should have no problem finding Mennonite support. The issues on which the coalition lobbies are conservation, wholesome food and fiber supply, wildlife habitat, and healthy rural communities. This is not to say that no Mennonites are involved in sustainable agriculture. In fact, my e-mail request for names of

persons involved in environmental activity yielded the name of a very
active environmentalist, Wayne Teel of Eastern Mennonite University.
He is active in the Sustainable Agriculture movement and is an advisor
to the Mennonite Central Committee on agriculture and environmental
issues.

A glimpse of some national voting statistics from the first session of
the 104th Congress (1995) will suffice to show where Mennonite political quietness has led (see table). The twenty congressional districts with
the largest concentrations of Mennonites are shown in descending order, with the League of Conservation Voters (LCV) rating of their representatives' voting records on environmental issues. The LCV tracked
the voting records of all representatives on a variety of issues that were
judged to be environmentally important. Twelve bills were used to make
these judgments. One hundred is a perfect score, and zero means that the
representative always voted wrongly on bills that were judged to be environmentally sensitive.[24]

It is possible, of course, to discount statistical studies of this nature.
Possibly the environmentally sensitive bills were mostly bad bills. If that
were the case, however, the LCV would have advised voting against the
bad bills, and a negative vote would have been counted as a positive value.
Perhaps more important than the raw scores is the fact that, as a group,
those representing areas most heavily populated with Mennonites
achieved an average score of 23.8, while the national average was 43.
Thus, even though the 104th Congress as a whole was judged by the
LCV to be profoundly unfriendly to the environment, the representatives
with the largest Mennonite populations in their districts scored just over
half as well as the (abysmal) average.

Two more aspects of this issue are even more disturbing than the
apparent lack of political effectiveness just described:

1. Mennonites are not standing shoulder to shoulder with other
churches who are taking up the message of salvation for the world that
God loves. This year has produced astounding news regarding the massive movement that evangelical Christians are making toward environmental action on the basis of biblical perspectives and spiritual commitments. In April the Evangelical Environmental Network claimed that
one thousand local churches had joined that fourteen-month-old movement. It was cofounded by Ron Sider, a Brethren in Christ professor of

Mennonite Voting Records (1995–1996)

District	Mennonite Population	Representative	LCV[a] Rating
PA-16	19,967	Robert Walker	0
PA-17	13,135	George Gekas	0
KS-01	9,420	Pat Roberts	8
IN-03	7,780	Tim Roemer	46
KS-04	6,713	Todd Tiahrt	0
PA-08	6,333	Jim Greenwood	77
PA-09	5,667	Bud Shuster	0
OH-16	5,546	Ralph Regula	23
PA-13	5,381	John Fox	54
VA-06	4,754	Bob Goodlatte	15
CA-20	4,438	Calvin Dooley	31
CA-19	4,247	George Radanovich	0
PA-19	3,866	William Goodling	15
VA-10	3,542	Frank Wolf	38
PA-06	3,256	Tim Holden	38
IL-18	3,243	Ray LaHood	23
OK-06	3,222	Frank Lucas	0
IN-04	3,178	Mark Souder	8
SD-00	3,019	Tim Johnson	54
NE-01	2,668	Doug Bereuter	46

[a]League of Conservation Voters.

theology at Eastern Baptist Seminary and president of Evangelicals for Social Action, and Dr. Calvin DeWitt, professor of environmental studies at the University of Wisconsin–Madison and a lifelong environmental thinker and activist. The evangelical embrace of environmental causes has been a long time coming but is the most promising and hopeful reality on the horizon today. Will the Mennonites be found in the Noah movement, as it is affectionately called?

2. By staying out of politics, Mennonites are alienating themselves from their most natural soul friends in the political world, who desperately need to be supported in the struggles they have undertaken as spiritual commitments. Secretary of the Interior Bruce Babbitt has increas-

ingly returned to his childhood faith for spiritual resources to carry on his struggle to protect the environment from the threats of destruction by human predators. He unashamedly links his public policy and his own deep spiritual commitment. The church of his childhood, he says, never taught him any obligations toward nature, but outside the church "I always had a nagging instinct that the vast landscape was somehow sacred, and holy, and connected to me in a sense that my catechism ignored."[25] Now, as a Catholic, Babbitt reaches out to faith communities across the country, accepting his political work as a God-given mission to save the earth.

CONCLUSION

On April 11, 1996, Bruce Babbitt flew to Phoenix, Arizona, where he addressed a group of some two hundred editors of Protestant and Catholic publications. His address, "Leading America Closer to the Promise of God's Covenant," was basically an altar call for the religious community to reclaim God's promises as given in the Genesis creation story. Early in his speech he spoke of the encouragement he had received in October 1995 in the form of five different letters from church leaders representing Presbyterian, Methodist, Evangelical Lutheran, Jewish, and Mennonite faiths. All of them opposed a bill that would cripple the Endangered Species Act. All of them opposed the bill not on the basis of scientific or technical or agricultural or medicinal reasons, but for spiritual reasons.[26]

The statement, "Stewards in God's Creation," was adopted in September 1995 as the mission statement for the Mennonite Central Committee. It is a strong and comprehensive statement, theologically sound and biblically based. One glaring omission, however, betrays a profound Mennonite weakness. There is only one small sentence expressing the necessity to be involved in the legislative process.[27] Even this brief and fleeting reference is negative, saying only that negative legislation should be opposed. Nothing in the document would lead one to think that Mennonites are or should be intentionally involved in the political arena. The *More with Less Cookbook*, praised in the document, seems to have captured the imagination of church thinkers with greater power than imagining how the church might be engaging the "powers." The publishing of cookbooks is not to be demeaned. Responsible consumption is at the core of our efforts to save the earth. But, having done this one thing well, have we neglected to do other things just as needful?

The 1996 spring seminar offered by the Mennonite Central Committee (MCC), Washington office, focused on "Engaging the Powers." Wilma Ann Bailey, professor at Messiah College, made an effective case for Anabaptists to become "advocates." She worked her way carefully through Anabaptist history and biblical literature, making a case for advocacy—Abraham pleads with God on behalf of the people of Sodom; Moses is an advocate for the people of Israel not only vis-à-vis the powerful Egyptians, but also with God. Bailey traced advocacy through the biblical corpus, noting that "to have someone to speak on our behalves is at the core of our understanding of what Christ has done for us."[28]

What would it mean for Anabaptists and Mennonites to become advocates for the earth, which lies at our feet without a voice, seeking redress from the ravages that we, the human family, have visited upon it? Perhaps as a result of the spring seminar, Karl Shelly, a staff person at the Washington MCC office, wrote eloquently on advocacy for the poor.[29] He notes that Anabaptists have a good track record in the matter of "taking care of our own" but should become more aware of the needs of others as we go on our way "to Jairus' house." The aptness of this interpretation should not escape any of us who are concerned about the fate of the earth. We are on our way to other things, some good and some not so good. We are preoccupied with our careers, our families, our institutions. But on our way to somewhere else, the earth interrupts us with persistent pleading, as did the woman who came to Jesus with her problem of continual hemorrhages for twelve years. Jesus was supremely interruptible, even while on a legitimate journey to heal the daughter of Jairus.

The pleading of the earth will not cease nor grow fainter. Will Mennonites be curators in their own museum, protecting the spiritual artifacts of their glorious past, or will they invite others to partake in the vision of the renewed kingdom of God? The church from its very beginnings has struggled with this dichotomy.

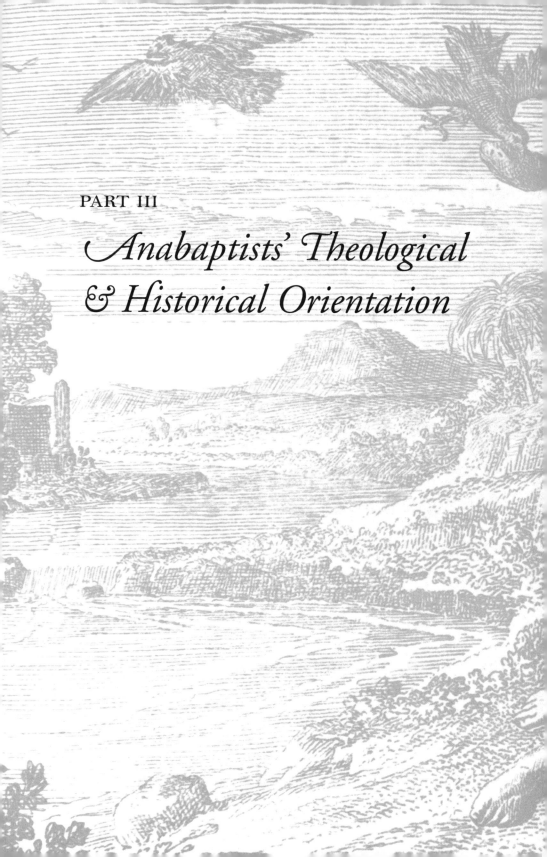

PART III

Anabaptists' Theological & Historical Orientation

CHAPTER 7

Creation, the Fall, and Humanity's Role in the Ecosystem

Theodore Hiebert

The biblical ideas of creation and fall and what these ideas might tell us about the human role in the ecosystem derive largely from the first three chapters in the Bible, the majestic account of creation in Genesis 1 and the familiar story of the Garden of Eden in Genesis 2 and 3. This is by no means all the Bible has to say about the creation of the world and about human disobedience, but these chapters have invariably provided the starting point for later reflection on these issues, and they have been referred to more frequently in recent ecological discussions than any other biblical texts.[1] So I use Genesis 1–3 as the basic text on the relevance of the biblical creation story in our age of environmental crisis. To explore its rich and diverse resources, I present for consideration three interpretations.

The first is the interpretation of the apostle Paul found in his correspondence to the early church—the New Testament reading. It is the interpretation we have grown up with and the one that most naturally comes to mind when we think of this text. The second is a historical reconstruction of the intentions of the authors of Genesis 1–3, the Priestly Writer, who is responsible for the account of creation in Genesis 1, and the Yahwist, from whom comes the Garden of Eden story in Genesis 2–3. I will call this the Old Testament reading. This interpretation combines data one finds in scholarly commentaries and my own reflections on the original setting and intent of these narratives. The third interpretation is modern, a response to the biblical text arrived at in an age of ecological crisis—the Anabaptist Creation Summit reading. This interpretation raises the issues I believe we should be discussing in our churches and preaching from our pulpits as we try to find a way of living in the world, a reading that preserves the integrity of creation as a whole.

THE NEW TESTAMENT READING

Let me begin, then, with the New Testament reading of Genesis 1–3, derived largely from the exegesis of the apostle Paul found in his correspondence to the Christians at Corinth and Rome (1 Cor. 15; Rom. 5, 8). As these letters show, the overriding concern Paul brought to the story of Adam and Eve was a concern about death, human mortality in particular, and this concern dominated his interpretation of the story. Paul's references to Adam and his act of disobedience indicate that Paul believed death was not part of the original creation. As Paul understood the story, human beings had been created sinless and immortal. Indeed, the entire creation was not initially subject to decay and death.

But Adam's act of disobedience, his eating of the forbidden fruit, introduced a catastrophic change into the nature of humanity and of the world of creation itself. As a result of Adam's sin, humans lost their immortality and became mortal, destined to die. "Sin came into the world through one man, and death came through sin," writes Paul to the Romans, "and so death spread to all because all have sinned" (5:12). As humans came under the powers of death through Adam's sin, so also did the wider world of nature. The creation (i.e., material creation apart from humans) was "subjected to futility," according to Paul (Rom. 8:20). It was placed under "bondage to decay," so that "the whole creation has been groaning in labor pains until now" (8:21–22). This drastic change in the nature of humanity and the created world, this subjection of them to the powers of death through Adam's sin, has been described in Christian theology as "the fall."

All this is the backdrop, according to Paul, for the real drama, the reintroduction of life into the world through Christ. For Paul, Christ was the new Adam or, better yet, the first Adam's polar opposite or reverse image. As Adam had brought death into the world, so had Christ conquered sin and death and made everything alive again. "If, because of the one man's trespass, death exercised dominion through that one," Paul tells the Romans, "much more surely will those who receive the abundance of grace and the free gift of righteousness exercise dominion in life through the one man, Jesus Christ" (5:17). Paul put it similarly to the Corinthians: "Since death came through a human being, the resurrection of the dead has also come through a human being; for as all die in Adam, so all will be made alive in Christ" (1 Cor. 15:21–22). And with the human race, all of nature will share in the redemption of Christ and be released from its subjection to death. "The creation waits with eager longing for the revealing of the children of God, . . . in hope that the creation itself will be set free from its bondage to decay and will obtain the freedom of the glory of the children of God" (Rom. 8:19, 21).

This reading of Genesis 1–3 as the story of a fall, in which death was introduced into the scheme of creation as a whole and into the human race in particular, is unique to Paul among New Testament writers, the majority of whom do not appeal to the Adam-Christ typology to express their understanding of Christ.[2] It is not quite correct, therefore, to call this reading of Genesis 1–3 the New Testament reading, as I have done, because it reflects the exegesis of Paul alone. Yet Paul's influence has been so great that his point of view has for all practical purposes become the Christian one, the one that probably first comes to mind when we turn to Genesis 1–3.

THE OLD TESTAMENT READING

Among the majority of Jewish interpreters through history and among many biblical scholars in the modern era, Genesis 1–3 has been interpreted very differently from Paul's reading. The Old Testament reading of Genesis 1–3 reflects the views of modern scholarship about the original setting and intentions within this text. Modern scholars have by and large attributed these chapters to two distinct authors, a Priestly Writer, responsible for the account of creation in a seven-day week in Gen. 1:1–2:4a, and the Yahwist, responsible for an alternate account of creation in the Garden of Eden in Gen. 2:4b–3:24. According to most modern scholars, neither of these authors was concerned primarily, as was Paul, with human mortality or with the conception of a catastrophic change in the scheme of creation brought about by human sin.

Let us begin with the account of the creation of the world in Genesis 1 by the Priestly Writer, or P. The overriding concern the Priestly Writer brought to the composition and transmission of this text was a concern for cultic order. His aim was to describe the establishment of "a stable cult in a stable cosmos."[3] To this end, P presents an account of creation that incorporates the origin of key elements of the religious institutions over which he presides. The most obvious of these is the sabbath, in the interests of which the entire seven-day scheme of creation is selected. By creating the world in six days and resting on the seventh, God establishes the sabbath at the beginning as part of the orders of the cosmos itself. Also established in P's creation account are the categories in nature upon which the dietary regulations of Priestly law were based. The realms of sky, sea, and land, into which the universe is carefully divided in the Priestly creation story, mark the boundaries by which clean and unclean foods are defined.[4] The repetitious style of the narrative itself may well reflect a liturgical setting, from which this literary masterpiece arose and whose worship it was intended to sustain and validate.

The role of the human in this cultically oriented scheme of creation has a distinctly priestly cast to it. Like the elite Priestly party in Israel, which saw itself as the mediator of divine presence to Israel and to the larger world alike, so the human in Genesis 1 is given a premier position in the entire created order. Alone among all forms of life, humans are made in God's image (1:26–27). By describing humans as created in the divine image, the Priestly Writer intended to define humanity as God's representative on Earth. In this representative capacity, humans are granted authority over the entire animal kingdom, the kind of authority by which ancient Near Eastern kings ruled their subjects (1:26, 28). And they are commissioned to subdue the earth, to place it into subjection under their power and control (1:28). Furthermore, humanity, the earth's governing species, is charged to reproduce, to become numerous, and to fill the earth (1:28). The human being in Genesis 1 is thus a powerful priestly figure, an intermediary between God and nature authorized to oversee and administer God's creation.

The Priestly image of creation and of the human place within it is quite different from the one later held by the apostle Paul. As Paul later also believed, the Priestly Writer thought the natural world was originally created good. But for P, the world's goodness never placed it outside the realm of death and decay. The Priestly emphasis in his description of the world's first vegetation is on its seed, the agent that ensures the continuance of plants in the ongoing cycle of life and death. And P's designation of humans as created in God's image does not identify them as originally immortal or in possession of a soul through which they could transcend the realm of death. The divine image, as I have already noted, gives humans a unique function in the world, not a unique essence or nature.[5] P's priesthood heroes got fairly far down the road to immortality, Methuselah in particular, who holds the record at 969 years. But they all died. And this death was not considered by the Priestly Writer as the result of any catastrophic change brought about in creation by human disobedience. The Priestly account of Israel's great ancestors— Adam, Noah, Abraham, and so on—does not treat them as fallen figures but as righteous heroes of Israelite history, priestly figures who mediate covenants between God and Israel. But this brings us to the second account of creation, the story of the Garden of Eden in Genesis 2 and 3, which the Priestly Writer incorporated into his primeval history and with which the notion of a drastic fall has been associated.

The story of the Garden of Eden, with its alternate version of the creation of the world, contains older Israelite traditions that modern scholars have customarily attributed to an anonymous author they have called the Yahwist, or J, who, unlike the Priestly Writer, prefers to refer to God in this early period of history by his personal name, Yahweh. The overriding concern that the Yahwist brought to his account of creation— and I am presenting a somewhat unique position on the Yahwist's primeval material—was a concern for agricultural productivity. The Yahwist's primary aim in the story of the Garden of Eden and in his stories of Cain and Abel and of the flood was to describe the origin of the relationship between the farmer and his land and to show how this relationship came to be defined and stabilized during the primeval era.[6]

The agricultural concerns that dominate the Yahwist's account of creation in the Eden narrative can be seen in his description of God's first creative acts. The world of nature that comes into existence at the beginning of this narrative is described as an irrigated orchard that God plants in the soil just as a farmer would (2:8). In this creation account humans are made to be farmers. The first man *'ādām*, is made out of arable soil, *'ădāmâ*, which establishes in the nature of things an innate and essential connection between humans and topsoil, a connection reflected even in the similarity of their names (2:7). And the task God assigns this first human is the cultivation of the soil from which he was made and the care of the orchard in which he is placed (2:8, 15). Although the archetypal human in Genesis 1 is a priestly figure, the archetypal human in Genesis 2 is the ordinary farmer.

From this point on, the real drama in the Yahwist's primeval traditions hinges on the relationship between the morality of the farmer and the productivity of the soil. This relationship is the predominant concern at the conclusion of each of the Yahwist's primeval stories—the Garden of Eden, Cain and Abel, and the great flood. God's initial plan, as it emerges in these narratives, is to link the land's productivity directly to human morality. Righteousness equals fertility, unrighteousness equals sterility. When the first man disobeys, God curses the arable soil on his account, making farming a more difficult struggle to produce a successful harvest (3:17–19). When his son Cain disobeys, God renders arable soil totally unresponsive to Cain's cultivation of it, forcing him off his land and into the arid wilderness (4:12). When God recognizes that the human race continues to devise nothing but evil deeds, he wipes out agricultural production completely, destroying the land, all of its produce, and all of its farmers except one, Noah, the only man the Yahwist calls righteous in the primeval age. On account of this catastrophe, God realizes that the creation is unsustainable if the land's productivity is linked directly to human morality. For the new age, therefore, he grants nature a new independent status, so that the land's productivity, the agricultural seasons and their harvests, and thereby the security of the farmer would be ensured regardless of the evil predilections of the human heart. "I won't curse arable soil again on account of humans," God says to himself after the flood, "because they are incurably evil ... As long as the earth lasts, seed time and grain harvest ... summer harvest and autumn harvest ... will not cease" (8:21–22).

In contrast to the powerful priestly role in creation assigned to humanity in Genesis 1, the Yahwist's archetypal farmer holds a much more modest position within the world of nature. The priestly human was differentiated from other animate life by being created in God's image, but the Yahwist's farmer is identified with the animals by being created out of the same topsoil from which they were made (2:7, 19). Whereas the priestly human is commanded to subdue the earth, the Yahwist's farmer is charged to cultivate the ground. In Hebrew, these verbs are precise opposites. Subdue, *kābaš*, means to master, to place the land under human control. Cultivate, *'ābad*, means literally to serve, to place human labor under the land's control. Thus, while the priestly human has a managerial role within the natural world, the Yahwist's farmer is more of an equal member of the community of life and a servant of nature's processes.

The Yahwist's narrative of the Garden of Eden in Genesis 2–3 is, of course, the primary Old Testament text upon which Paul based his theology of the fall of humanity and of creation as a prelude to Christ's redemption. Indeed, the presence of death and decay in humanity and nature, which so concerned Paul, is an obvious theme in this text, but the Yahwist J handles this theme in very different ways than Paul was later to do. For J, human mortality was not the overriding concern of the Eden narrative, nor was it linked to human sin. J considered death a fact of existence, built into the created order when the first human was made from the soil to which humans were subsequently destined to return (2:7, 19). Immortality—the kind of eternal life possessed by God—was available, like knowledge, only from the fruit of one of Eden's trees, a tree placed off-limits to humans when the first couple was sent away from the garden (3:22–24). Death was thus considered part of the human condition, which differentiated humans from God, rather than the result of human sin. Disobedience in the garden was punished by other sentences: difficulties in conception and childbirth for the woman and difficulties in agricultural production for the man.

Paul's view that human sin introduced a kind of fall into the broader creation outside of humanity comes somewhat closer to J's point of view and major concern than to P's. In each of the Yahwist's primeval narratives, the land suffers because of human immorality. But this ultimately disastrous relationship between human sin and nature's sterility is a temporary imbalance rectified through the flood hero Noah. When Noah is born, he is identified as the one who will bring the human race relief from the curse on the land (5:29). The only righteous man in his generation (7:1), Noah was saved from the flood, and when he made an offering to

God upon exiting the ark, God made the pledge to which I have already referred, to lift the curse and guarantee the stability of the agricultural year as long as the earth lasts. For the Yahwist, Noah was the new Adam, the man in whom God began a new era in which nature would be restored to its precurse status and thereby provide a stable context for human life for all time to come.

THE ANABAPTIST CREATION SUMMIT READING

Having described three biblical views of creation, the fall, and humanity's role in the ecosystem—the Yahwist's agricultural view, the Priestly Writer's cultic view, and Paul's later Christological view—I introduce one more perspective, our own twentieth-century context in which the human environment has been dangerously imperiled by human irresponsibility. And I offer a few reflections on how the topic under consideration looks from here, commenting in particular upon the resources within these biblical views for shaping a responsible contemporary approach to creation.

Let me begin with the concepts of creation and fall, that is, with the status of the world of nature as a whole. The contemporary biologist or ecologist would take nature as a kind of given, a complex and magnificent system of relations between organisms and their environments that evolved over millions of years. When we talk about caring for the earth, we mean preserving as far as possible the health of this complex system.

In light of this contemporary view of creation, I find Paul's reading of Genesis, by which nature is regarded as fallen and in need of redemption, a particular problem. In the first place, Paul's view that death entered nature as a kind of foreign element on account of human sin does not seem to be an accurate description of a natural system whose very existence is predicated upon the ongoing cycle of life and death. In the second place, Paul's view leads to perspectives that can easily run counter to caring for this ecosystem. It stands behind the kind of apocalyptic thinking represented by former U.S. Secretary of the Interior James Watt, who defended the exploitation of the earth's resources on the presumption that this old earth would pass away to be replaced by a new creation. Paul's view invites us not to accept the nature we know as a standard or rule to which we must attune our behavior. We are prone to think of it instead as in need of change, in need of redemption. And when we speak of such changes, they are usually for human benefit, as when the lion lies down with the lamb or the wilderness comes to life.[7]

As descriptions of the status of nature as we know it, are not the Old Testament perspectives on creation a more appropriate starting point for a modern ecological theology? Neither the Priestly nor the Yahwistic account of creation is composed according to the principles of modern science, of course. These authors lived before the era of modern science, and each, moreover, brought a special social agenda to his creation narratives, whether priestly or agrarian. Yet both accept nature fundamentally as it is, regarding it as reflecting the patterns and orders God intended at creation. The Priestly Writer is particularly emphatic about this, punctuating his account of the creation with the sevenfold refrain, "And God saw that it was good." For the Yahwist, a particular element of nature, arable land, does suffer from the curse in the primeval age. But after the flood, the curse on the soil was lifted and the stable rhythm of agricultural productivity known to the Yahwist was guaranteed for all time. Both authors thus regard the scheme of creation with which they were familiar as a sacred order designed by God to sustain life and to which human behavior should be attuned.

Now let me turn to the issue of humanity's role within the ecosystem and reflect briefly on the two sides of human behavior—responsibility and irresponsibility, righteousness and sin—as these are described by biblical writers. In the texts under consideration, the Yahwist and Paul in his interpretation of the Yahwist show a particular sensitivity to the relationship between human morality and creation. In his primeval narratives, J connects the consequences of human disobedience and sin in society primarily with the larger natural world his humans inhabit. Human sin results in the land's sterility. Paul pushes this concept to cosmic dimensions, blaming death in all of nature on human sin. In an era when our government and business leaders do not commit crimes but only make errors in judgment, in an era when Louisiana-Pacific can run an ad describing "how precious our God-given resources are, and how important it is never to waste them" while piling up over forty thousand violations of the Clean Water Act, we need to recover in some fashion the biblical recognition of human sin and its consequences to our natural habitat.[8]

But this recovery cannot be a simple, straightforward appropriation of the biblical point of view. The rigid biblical equation between a natural disaster—be it disease, epidemic, famine, earthquake, flood, or thorns and thistles—and human sin has many hazards that must be successfully negotiated. Most natural powers—from the microorganisms that cause disease to the huge tectonic forces behind earthquakes—have nothing to do with human morality one way or another. Failure to recognize this leads to hubris, by which we think we are more powerful or important in the ecosystem than we actually are, or to unjust and uncompassionate judgment of those who experience tragedies that they did nothing to cause. To understand human culpability properly, we need to develop a clear sense of both the power and the limits of human behavior, a sense of precisely where we stand in the ecosystem.

This leads to an observation about where biblical authors actually position humanity within the ecosystem and what responsibilities are entailed in this role. In this regard the Priestly Writer and the Yahwist offer us an interesting contrast. According to P in Genesis 1, humans have been assigned by God to a powerful, managerial role in creation. They have been invested with authority and control over the earth and its life and given responsibility to administer it for their divine sovereign. From this priestly image of the human has come the modern concept of the human as steward of creation, surely the most common characterization of the human role in nature among Christians today.

According to the Yahwist in Genesis 2, by contrast, God has assigned humans a much more modest position in the ecosystem. Made out of topsoil, as was all other life with which humans share a common nature, the first human was commissioned to cultivate, literally "to serve," the soil. Thus, humanity was instructed to place itself in the service of the land and its vegetation, upon which human survival depended. To distinguish the Yahwist's human from the priestly human, I would call him the servant of creation, taking this designation from the verb by which God defined humanity's role in the garden.

In these two biblical views of humanity's role in the ecosystem, we have in a nutshell the paradox of human existence. On the one hand, we know ourselves to be—among all species of life—uniquely powerful and inventive, uniquely able to manipulate and alter nature for our ends, a kind of priestly presider over creation. On the other hand, we know that we are not at all in control—the theme, by the way, of an editorial in the *Los Angeles Times* after the recent earthquake. Our very survival is dependent on biological, chemical, and geological processes we hardly understand and certainly cannot control. And this human paradox has only been intensified by modern science, which has given us the power to manipulate nature as never before and taught us at the same time what a tiny, insignificant, and expendable event human life is within the scope of evolutionary time.

There is something to be said for maintaining a healthy relationship between both of these images, explanatory as they are of our paradoxical position in the world. Yet I would like to urge, in conclusion, that we develop a new appreciation for the human as a servant of creation, an image of the human that resonates strongly with the Anabaptist conception of the simple life. The practitioners of a new ecological way of life who have been most persuasive argue that we have to give up our ideas that we have enough knowledge and power to manage creation and recognize that we must learn to conform our behavior to nature's own processes and demands if we hope to survive. To illustrate, I conclude with a few sentences from Wendell Berry, who, like the Yahwist, writes as a small farmer.

> That humans are small within the Creation is an ancient perception, represented often enough in art that it must be supposed to have an elemental importance ... The message seems essentially that of the voice out of the whirlwind in the book of Job: the Creation is bounteous and mysterious, and humanity is only a part of it—not its equal, much less its master ... Creation provides a place for humans, but it is greater than humanity and within it even great men are small. Such humility is the consequence of an accurate insight, ecological in its bearing, not a pious deference to "spiritual" value ... We will have either to live within our limits, within the human definition, or not live at all.[9]

CHAPTER 8

The New Testament and the Environment: Toward a Christology for the Cosmos

Dorothy Jean Weaver

In his 1979 essay "Biblical Views of Nature," John Austin Baker concludes that, "in contrast with the Hebrew Scriptures, the New Testament has relatively little to say about nature." There is truth in Baker's assessment. As he puts it, "there is nothing in the New Testament ... to parallel the large collections of 'observations on life and the world-order,' which we call the wisdom literature of the Hebrew Scriptures, or its extensive range of liturgical poetry, or the detailed corpus of its laws on what we would regard as secular matters. The very types of material, therefore, in which an attitude to nature might be most likely to be reflected are precisely those missing from the New Testament."[1] Raymond C. Van Leeuwen summarizes the same situation in positive terms: "The New Testament is a small book with an infinitely important but very limited agenda: to proclaim the gospel of God's grace revealed in the life, death, and resurrection of Jesus Christ; to teach and guide the infant church of Christ, newborn at Pentecost, in light of the gospel; and to point to the final renewal of all things."[2]

Accordingly, the central problem with our search to find New Testament perspectives on environmental issues lies in the fact that the New Testament is not asking questions about the environment. Further, the environmental issues with which we are concerned belong to today's world, the industrialized global village of the twenty-first century, a world vastly different from that of the New Testament. The New Testament writers do not and cannot answer our environmental questions for the simple reason that our issues did not exist in their world. As Van Leeuwen points out, "It is no longer the first century A.D. Without much thought, one can see that the world has changed since the days Jesus walked in sandals and a robe. There are no New Testament texts directly offering guidance and commentary on nuclear war and waste, genetic engineering, wetlands, ozone depletion, rain forest burning, garbage disposal, or MTV."[3]

Where does this leave us in our search for a faith response to environmental issues, since the New Testament writers have questions different from ours today? Even if they are headed toward a different goal, in their very proclamation of "the good news of Jesus Christ" they do have a great deal to say about the world that Jesus graced with his life, death, and resurrection. As a result there is a profound linkage between "the good news of Jesus Christ" and the questions we now ask about our environment. According to Gordon Zerbe, "the New Testament projects a vision of the kingdom of God that is full of implications for a Christian environmental ethic."[4]

CHRIST AND CREATION

The central motif throughout New Testament writings is that of the "kingdom of God." And the overarching message of the New Testament lies in the "good news of the kingdom." As the New Testament writers proclaim it, the "good news" is that "the kingdom of God has come near" to humankind (Mark 1:15) in the life, death, and resurrection of Jesus of Nazareth. It follows that the "good news of the kingdom" is integrally linked to our environmental questions as the whole is to the part. The God whose sovereign rule Jesus proclaims and to whom Jesus stands in the relationship of "Son" to "Father" is, by Jesus' own designation, none other than "Lord of heaven and earth" (Matt. 11:25), the Creator God identified with "the creation that [He] created" (Mark 13:19).[5]

And what the Jesus of the Gospels proclaims with reference to "God as Creator" finds multiple echoes throughout the New Testament. God is the "Creator . . . [whose] eternal power and divine nature . . . have been understood and seen—[ever since the creation of the world]—through the things he has made" (Rom. 1:20, 25) and the one who accordingly has proprietary rights to "the earth and all its fullness" (1 Cor. 10:26).[6] Other New Testament writings emphasize the comprehensiveness of God's creative work (Acts 4:24; Heb. 2:10; 2 Pet. 3:5; Rev. 4:11), specific details of God's creative work (2 Cor. 4:6; Heb. 1:10; 4:3–4; 11:3; 2 Pet. 3:5), and God's activity in the creation of humankind (Mark 10:6; Acts 17:25–26; 1 Cor. 11:8–9; 15:47; James 1:18). But the motif of "God as Creator," prominent and significant as it is, does no more than identify the New Testament writers as members of the Jewish faith community who are reciting the "mighty acts of God" (Deut. 3:24; Ps. 145:4) as recounted in their Scriptures. For the early Christians there is nothing essentially new about the New Testament confession that "God is Creator."

What is strikingly new, however, is the New Testament witness to the agent through whom God has enacted the creation of all things: The God who has acted to bring into being all things that exist has done so precisely through the agency of Jesus Christ. While they express this faith in a variety of ways, New Testament writers both early and late share a common Christological view of creation.[7] Christ is the one "through whom are all things and through whom we exist" (1 Cor. 8:6), the one "through whom God also created the worlds" (Heb. 1:2), and the one "in [whom] all things in heaven and on earth were created" (Col. 1:16).

Nor is this all. If Christ is the agent through whom God has created all things, then this is because Christ was present and active "in the beginning" when God was at work. The "Word" is "in the beginning with God" (John 1:1–2). Christ is "the firstborn of all creation" (Col. 1:15) and the "origin [or beginning] of God's creation" (Rev. 3:14). Accordingly, the Jewish "theology of creation," clearly visible elsewhere within the New Testament, has now in effect been transformed, point by point, into an early Christian "Christology of creation."

And this is no minor transformation. The presence of the "creation" motif within the New Testament, in both its theological and its Christological formulations, has profound implications for our questions concerning the environment. God the Sovereign Ruler is by the same token God the Creator. And Christ the Redeemer is none other than Christ the Agent of Creation. Accordingly, to acknowledge God's rule and God's redemption is to stand in reverence before God the Creator, to submit our lives to the claims of Christ the Agent of Creation, and to treat with utmost care and respect all that which God has brought into being through Christ.

INCARNATION AND THE NATURAL WORLD

A second prominent New Testament motif with implications for the environment is that of incarnation. In contrast to the creation motif, which bears witness above all to the transcendent power of God, the incarnation motif bears witness to God's immanence in the created order and God's solidarity with humankind through the person of Jesus Christ.

As the New Testament writers bear witness, it is precisely Christ the Agent of Creation who takes on human form and becomes one with the human community. John the Gospel Writer describes this event in terms of "the Word" who "was in the beginning with God" (John 1:2) and who "was God" (John 1:1). As John puts it, "the Word became flesh and lived among us, and we have seen his glory" (John 1:14).

The logic of this incarnational theology moves in two directions. On the one hand the motif of incarnation moves in a downward direction and points us to that which Loren Wilkinson identifies as "the central paradox of the gospel," namely "that the Creator God, the Lord of the universe, has become flesh, has entered the human condition, has submitted to the limits of time and place, and has become a living human being along with us." This central paradox is the driving theological force behind the Gospel of John.[8] Following his programmatic statement of 1:14 that "the Word became flesh," John spends the remainder of his narrative depicting Jesus as the one "from above" (3:31; 8:23), the one who has "come down from heaven" (3:13, 31; 6:31, 32, 33, 38, 41, 42, 50, 51, 58) in order to reveal God to humankind through the earthly, material "stuff of life." There can be no question about the genuine "humanity" and "earthliness" of the "one come down from heaven" as John portrays him. In an extraordinary act of divine solidarity and downward mobility, God has come to live among humankind as a mortal human being.

On the other hand, the incarnation motif in the Gospel of John moves in an upward direction, pointing us to the capacity of earthly substances to mediate the presence of the divine. If John seeks to show us through his gospel that God has taken on human flesh in the person of Jesus, he also seeks to show us that material reality is likewise, in John Habgood's words, "capable of bearing the image of the divine."[9]

This becomes evident in the "signs" that Jesus performs.[10] Ordinary earthly substances take on extraordinary qualities at Jesus' initiative. Water is transformed into wine (2:1–11); five barley loaves and two fish become a feast for five thousand people (6:1–15); a turbulent sea provides stable footing for Jesus to walk to his disciples in their boat (6:16–25); mud from the ground becomes the healing ointment for the eyes of a blind man (9:1–7); and a previously barren sea instantaneously yields up 153 fish (21:1–14). In John's gospel, material reality—the everyday stuff of life—becomes a window through which the viewer catches recurring glimpses of the glory of God.

If, then, earthly stuff of Jesus' "signs" opens "windows" onto divine reality, the same is true for the earthly analogies of Jesus' "I am" sayings, the parabolic self-designations by which Jesus interprets his "signs" and proclaims his identity. Jesus, who asks a drink from a Samaritan woman, offers her in return "living water" (4:10; 4:11) that is "gushing up to eternal life" (4:14). Jesus, who has created a feast for the masses from a handful of loaves and fish, announces that he himself is "the bread of life" (6:35, 48) and "the living bread that came down from heaven" (6:51). Jesus, who gives sight to a blind man, proclaims himself "the light of the world" (8:12; 9:5). Elsewhere Jesus identifies himself as "the door of the sheep" (10:7) and "the true vine" (15:1; 15:5). In John's view ordinary matter and natural phenomena—the stuff of touch, taste, sound, sight, and smell—become the occasions of epiphany, the tangible, sensory means by which Jesus reveals intangible divine reality to a humanity who has "never seen God" (cf. 1:18).

Further, it is in the corporate activities to which Jesus calls his followers in the Gospel of John, those things which for us have become the liturgical practices of the church, that Jesus' use of the tangible world as a "window" onto divine reality becomes most explicit. In the "eucharistic" discourse following the feeding of the five thousand (John 6:22–58), Jesus proclaims himself to be the "bread of life" (6:35, 48) and promises his disciples that, as they "eat the flesh of the Son of Man and drink his blood" (6:53), they will "abide" in him (6:56), "[be raised up by him] on the last day" (6:54), and have "eternal life" (6:54). From John's perspective it is as the followers of Jesus partake of the *tangible earthly substances* of bread and wine that they are nourished into *spiritual life* by the "true food" of Jesus' flesh and the "true drink" of Jesus' blood (6:55), which have been "given for the life of the world" (6:51).[11]

The Synoptic Gospel writers communicate the reality of Jesus' "incarnation" in their own way through their accounts of Jesus' "transfiguration" (Mark 9:2–8 parr.). In these accounts the earthly Jesus is "transfigured" (Mark 9:2; Matt. 17:2) and his face "shines like the sun" (Matt. 17:2) with what can only be viewed as the very glory of God. And the same glory that transfigures Jesus himself transfigures his clothes as well into "dazzling white" garments (Mark 9:3). As Elizabeth Briere notes, "through the human person [of Christ], the rest of material creation also becomes transparent to God, thus realizing the potential for which it was created."[12]

The implications of incarnation theology for our approach to environmental issues are simple and at the same time profound. Material creation, the everyday ordinary "stuff of life," is of utmost significance in the divine scheme of things. This is so not only because God *through Jesus Christ* has created this "matter" in the first place, but even more crucially because God *in Jesus Christ* has chosen to enter into the realm of created matter and take on the "mortal flesh" of humankind.[13] In Briere's words, "Because God himself has become matter, 'I will not cease to honour the matter which brought about my salvation,' as Saint John of Damascus writes."[14]

CREATION AS GOD INTENDED

The New Testament writers bear sturdy witness to the role of Christ as Agent of Creation and the identity of Christ as the Incarnate One. This witness signals in broad and fundamental terms the positive significance of created matter. The writers likewise offer a closer, more detailed portrayal of the created world order. Both in the teachings and ministry of Jesus and elsewhere throughout the New Testament we find a portrayal of "creation as God intended it."

To begin with, the New Testament writers offer persistent witness to the continuing acts of God in sustaining creation. God is the one who "makes his sun rise" and "sends rain" on the earth (Matt. 5:45; cf. James 5:17–18); who "clothes the grass of the field" (Matt. 6:30; cf. 6:28–29) and "gives the growth" to things that are "planted" and "watered" (1 Cor. 3:6–8);[15] who "feeds the birds of the air" (Matt. 6:26) and sees to it that "not one sparrow falls to the ground" unnoticed (Matt. 10:29); who "supplies seed to the sower and bread for food" (2 Cor. 9:10); who provided "manna in the wilderness" for the Jewish "ancestors" to eat (John 6:31). And in the parabolic language of the New Testament, God is portrayed in numerous "caretaker" roles vis-à-vis the earth.[16]

The New Testament writings likewise point to the crucial role of Jesus in sustaining creation. He is the one "in [whom] all things hold together" (Col. 1:17) and the one "[who] sustains all things by his powerful word" (Heb. 1:3). And when faced with a challenge to his "lifegiving" ministry of healing (5:21, 25–29), the Johannine Jesus appeals to the ongoing work of the Creator God as the source and the authority of his own activities: "My Father is still working, and I also am working" (5:17).[17] Since God the Creator, Jesus' "Father," is still active in creation, Jesus, God's "Son," is likewise at work in a healing ministry of his own that parallels the "lifegiving" acts of God and draws its power from God the Giver of Life (5:21, 26).

Accordingly, as John sees it, Jesus' entire ministry is to be viewed as the continuation of God's original act of creation.[18] On the one hand, Jesus carries on God's creative work through the "acts of power"/"signs" by which he heals sick people (Matt. 4:23–25), casts out evil spirits (Mark 5:1–20), takes authority over the turbulent forces of nature (Mark 4:35–41; 6:45–52), and raises the dead to life (John 11:1–44).[19] Jesus' creative work is clearly visible on those occasions when his "acts of power"/"signs" provide an abundance of food for hungry crowds in the wilderness (Mark 6:31–44; 8:1–10), huge vats of wine for thirsty wedding guests (John 2:1–11), and overflowing nets of fish for unsuccessful fishermen (Luke 5:1–11; John 21:1–14). In the metaphorical language of the "caretaker of creation," Jesus describes himself as the "[sower] who sows the good seed" (Matt. 13:37), the "good shepherd ... [who] lays down his life for the sheep" (John 10:11, 14–15), and the "hen [who] gathers her brood under her wings" (Matt. 23:37).

JESUS' FOLLOWERS AND GOD'S CREATIVE WORK

On the other hand, Jesus calls his disciples to join him in carrying forward God's creative work. To a question from his disciples about a blind man, Jesus responds (John 9:4): "Neither this man nor his parents sinned; he was born blind so that God's works might be revealed in him. We must work the works of him who sent me while it is day. Night is coming when no one can work." Here again Jesus locates the source of his ministry in the activities of God the Creator, the one who has "sent" him to his earthly assignment. It is none other than the creative "works" of God which Jesus will "reveal" as he brings "light" into the "world" of the blind man (cf. Gen. 1:3; John 1:4–5, 7–9). And with his pointedly plural address ("*We* must work the works of him who sent *me*"), Jesus commissions his disciples to join him in this task of carrying forward God's creative work— salvation of humanity and its sustaining milieu, the earth.

For their part, the New Testament writers offer numerous clues to the nature of this task. The teachings of Jesus and the early Christians are filled with parables and analogies drawn from the everyday world of first-century Mediterranean agriculture and animal husbandry. These parables and analogies provide a vivid and detailed portrait of "earthkeeping" in the first-century Mediterranean world.

The world of these sayings is one in which people work closely and intensively with the earth to ensure its "fruitfulness." We see them "plowing" (Luke 17:7; 1 Cor. 9:10), "sowing seed" (Mark 4:3–4; Matt. 13:24; 1 Cor. 15:37), "watering" (1 Cor. 3:6–8),[20] "digging" (Luke 13:8), "pruning" (John 15:2), and "grafting" (Rom. 11:17, 19, 23, 24). And when the plants grow and "bear fruit" (John 12:24), we see people "mowing" (James 5:4), "reaping" (Matt. 9:37–38; John 4:35–38; Gal. 6:9; James 5:4), "threshing" (1 Cor. 9:10), "winnowing" (Matt. 3:12; Luke 3:17), and "gathering into barns" (Luke 12:18).

The "earthkeeping" implications of this agricultural imagery are clear. The earth is intended to be a "fruitful" place. It is as plants "bear fruit" that they fulfill the signal purpose of their existence (Mark 4:8).[21] Accordingly, for Christians to be engaged along with Jesus in "working the [creative] works of God" involves working together with the earth to ensure that plants "bear" the "fruit" they are intended to produce.[22]

The New Testament texts likewise depict a world in which humans serve as caretakers for other living creatures. People "tend"/"shepherd" their flocks of sheep (Matt. 2:6; Luke 17:7; John 21:16; 1 Cor. 9:7; Rev. 7:17), "call their sheep by name and lead them out . . . [to] find pasture" (John 10:3, 9), "guide them to springs of . . . water" (Rev. 7:17), protect them from being "scattered" (cf. Mark 14:27; Matt. 9:36), "keep watch" over them (Luke 2:8), provide "sheepfolds" for their safekeeping (John 10:1, 16), and search for them when they are "lost" (Luke 15:4) or have "gone astray" (Matt. 18:12). Similarly, people "give water" to their donkeys (Luke 13:15), "herd" their swine (Mark 5:14), and "feed" them with "pods" (cf. Luke 15:15–16). The "earthkeeping" envisioned by the New Testament writers clearly involves care for God's creatures, just as it involves care for the earth and the plants.

CREATION SERVING HUMANKIND

If the New Testament writings depict humans nurturing the created world around them—earth, plants, and living creatures alike—they also portray the created world, in reciprocal fashion, assisting and nurturing humankind. People "tame every species of beast and bird, of reptile and sea creature" (James 3:7)[23] and draw on their services in a variety of ways. Ordinary people employ donkeys as a means of transportation (Matt. 21:2, 7), and military personnel ride horses (Acts 23:24).[24] People use oxen for plowing (Luke 14:19) and for "treading out the grain" (1 Cor. 9:9; 1 Tim. 5:18).

Similarly, humans receive physical nourishment from the fruits and the creatures of the created world. People "eat the fruit of the vineyard" (cf. 1 Cor. 9:7) and "drink the fruit of the vine" (Mark 14:25). They eat bread baked from "[wheat] flour" (Luke 13:21), figs picked from their trees (Mark 11:12–13), eggs from their chickens (Luke 11:12), herbs from their gardens (Matt. 23:23), and fish drawn from their waters (Luke 11:11; Mark 6:41). They drink water from their wells (John 4:1–15) and milk from their flocks (1 Cor. 9:7). They feast on the meat of "fatted calves" (Luke 15:23, 27, 30) and goats (Luke 15:29) and celebrate the Passover by eating the "paschal lamb" (Mark 14:12; 1 Cor. 5:7).

Accordingly, "creation as God intended it" is not only a "fruitful" and sustaining environment, but a fundamentally interdependent realm in which humankind and the rest of the created order *mutually* and *necessarily* support and nurture each other. As the New Testament writings bear witness, the well-being of all creation—earth, plants, living creatures, and human beings—ultimately depends on the well-being of each of the constituent groups. The "earthy" parables and analogies found throughout the New Testament offer a macrocosmic parallel to the microcosmic "body theology" of the Pauline and Deuteropauline epistles (1 Cor. 12:12–26; Eph. 4:1–16; Col. 2:19). To rephrase Paul's words, "if one element of creation suffers, all suffer together with it; if one element of creation is honored, all rejoice together with it" (cf. 1 Cor. 12:26).

"Creation as God intended it" is a place of physical engagement and enjoyment, a hands-on world whose many riches human beings are invited to "handle, taste, and touch" with confidence (cf. Col. 2:21) and to "partake with thankfulness" (1 Cor. 10:30). Precisely because "the earth and its fullness are the Lord's" (1 Cor. 10:26), the riches of the earth represent God's "good gifts" to humankind (Luke 11:13), gifts intended to be nurtured, experienced, and enjoyed.

Just as importantly, however, "creation as God intended it" is a place in which humans share with each other the riches of the created order. Jesus depicts harsh judgment falling on those who "build larger barns" so that they can hoard their abundance (Luke 12:21) and those who "feast sumptuously" while their neighbors go hungry (Luke 16:19–31). He calls the wealthy to "sell what they own and give the money to the poor" so that they can then "come and follow" him (Mark 10:17–30). And he proclaims ultimate "blessedness" to those who "feed the hungry, give drink to the thirsty, welcome the stranger, clothe the naked, take care of the sick, and visit the imprisoned" (Matt. 25:31–46). Paul, for his part, challenges the Corinthians to share their material wealth with the Christians at Jerusalem in order to create "a fair balance between your present abundance and their need" (2 Cor. 8:13–14). And he assures the Corinthians that God will supply them with abundance precisely in order to enable their generosity (2 Cor. 9:8). "Creation as God intended it" is ultimately a place of "shared abundance."[25]

HUMAN SINFULNESS AND THE DEVASTATION
OF CREATION

While the New Testament writers forthrightly proclaim the goodness of "creation as God intended it," they by no means romanticize the
created world or downplay the harsh realities of life in that world. To
the contrary, the New Testament writers are solidly pragmatic about the
created order and realistic about the numerous life-threatening dangers
posed by this created order.

Undoubtedly, the most prominent and poignant illustration of this
New Testament realism about the created world order lies in the repeated
references to physical illness and demon possession. As the Gospels portray it, one of the central foci of Jesus' public ministry is the healing of
those who are sick. Jesus summarizes his ministry with a list of healing
activities (Matt. 11:5; Luke 7:22) and identifies the "coming of the kingdom of God" with his actions in "casting out demons by the Spirit/fin-
ger of God" (Matt. 12:28; Luke 11:20). He commissions his disciples to
carry forward his ministry as they "cure the sick, raise the dead, cleanse
the lepers, [and] cast out demons" (Matt. 10:8). For his part the apostle
Paul identifies death as "the last enemy to be destroyed" (1 Cor. 15:26),
while John the Revelator describes "a new heaven and a new earth" (Rev.
21:1) in which "death will be no more" (Rev. 21:4). As the New Testament writings make clear, human illness and death are evils that stand
in direct opposition to "creation as God intended it."

But human illness is not the only such breakdown of the created order. Sheep "go astray" (Matt. 18:12–13; 1 Pet. 2:25), get "lost" (Luke 15:4,
6), and are attacked and ravaged by wolves (cf. Matt. 7:15; Luke 10:3;
John 10:12). Sparrows "fall to the ground" and die (Matt. 10:29). Weeds
"choke" other plants so that they "yield nothing" and "their fruit does not
mature" (Mark 4:7, 19). Hail destroys plant life and injures people (cf.
Rev. 8:7; 11:19; 16:21). The heat of the sun "withers" plants (Matt. 13:6;
James 1:11) and "scorches" humans (Matt. 20:12; Luke 12:55). Drought
gives rise to famine (Luke 4:25; 15:14). Winds, rain, and flooding rivers
threaten some houses (Luke 6:48) and destroy others (Luke 6:49). Winds,
waves, and rough seas "swamp" and "batter" fishing boats (Mark 4:37;
6:48) and threaten humans with death by drowning (Mark 4:38; Matt.
14:30). Vicious storms on the ocean cause large ships to wreck and

threaten the lives of all on board (Acts 27:13–44). A catastrophic flood "deluges" the earth with water and "sweeps away" almost all human life (Matt. 24:37–39; 2 Pet. 2:5; 3:6). Earthquakes "split" cities, cause their "fall," and destroy their inhabitants (Rev. 11:13; 16:18–19).[26] Just as the created order is fruitful and nurturing on the one hand, it is threatening and destructive on the other.

In Rom. 8:18–23 Paul describes this reality and offers an explanation of it in terms of the cosmic plan of God. As Paul sees it, the creation has not yet arrived at its intended goal. He chooses an array of images, some positive and some negative, to drive home this reality. On the negative side, creation has been "subjected to futility" (v. 20) and presently experiences "bondage to decay" (v. 21). On the positive side, creation "waits with eager longing" (v. 19) and "groans in labor pains" (v. 22) in anticipation of glory that lies ahead but has not yet been attained.[27] But if creation has not yet arrived at its intended goal, neither has humankind. As Paul puts it, "we ourselves . . . groan inwardly while we wait for adoption, the redemption of our bodies" (v. 23). And it is this human, bodily "redemption"—otherwise described as "the revealing of the children of God" (v. 19)—on which "the whole creation" waits for its own release from "bondage to decay."

The overall import of Paul's assessment, accordingly, can hardly be mistaken. Not only do creation and humankind alike stand somewhere short of their intended goals, but their arrival at these goals is also integrally linked. "The whole of creation" cannot arrive at its goal apart from humankind, nor can humankind experience its own redemption apart from the rest of the created order. As Paul views it, humankind and the rest of creation together constitute a cosmic body whose members are interdependent upon each other and whose destinies are inextricably intertwined.[28]

What Paul encapsulates in his language about creation's "bondage to decay" John the Revelator paints in great detail and in vivid color in his letter to the churches of Asia Minor. It is here, more than in all other New Testament writings,[29] where we glimpse what seems to be an environmental message directed not only to the first-century Mediterranean world but also, in very immediate fashion, to the global village of the twenty-first century. In environmental terms the prophetic words of John the Revelator require astonishingly little interpretation to be understood in our context; their relevance to our world is hard to escape.

Here John offers his readers a panoramic scene of catastrophic cosmic breakdown that encompasses every element of the created order with its devastation. On the human level there is war and human slaughter (6:1–2, 3–4, 7–8, 9–11), along with steep inflation affecting basic commodities (6:5–6). People suffer from "foul and painful sores" (16:2) and "every kind of plague" (11:6). They experience drought (11:6), "flood" (12:15), the "fierce" and "scorching" heat of the sun (16:8–9), and "dried up" rivers (16:12). They die from "famine" (6:7–8; 18:8), "pestilence" (6:7–8; 18:8), "wild animals" (6:7–8), and water polluted by "wormwood" (8:10–11). Smoke "like the smoke of a great furnace" ascends and "darkens the sun and the air" with its pollution (9:2). Earthquakes "split cities" (16:18), cause them to "fall" (11:13), and decimate their populations (11:13).

This cosmic catastrophe likewise affects all the rest of the created order. For its part the earth experiences major devastation. Land masses are dislocated such that "every mountain and every island is removed from its place" (6:14). Major conflagrations "damage" and "burn up" much of the earth (7:3; 8:7). The rivers turn to "wormwood" (8:10–11), and the "waters" turn to "blood" (11:6). Living creatures in the sea die (8:9; 16:3), and ships are "destroyed" (8:9). The celestial bodies also participate in this breakdown of creation. The sun "becomes black as sackcloth" (6:12), the full moon "becomes like blood" (6:12), the stars "fall to the earth" (6:12), and "their light is darkened" (8:12). And the sky itself "vanishes like a scroll rolling itself up" (6:14).

But if John the Revelator takes great pains to depict the cosmic breakdown of the created order, he communicates with equal conviction the cause of this catastrophe: "The harvest of the earth is fully ripe" and must therefore be "reaped" (14:14–20).[30] In other words, it is "those who destroy [or corrupt] the earth" who will encounter their own ultimate "destruction" in the cosmic upheaval that John depicts (11:18; 19:2). As John views it, human sinfulness and failure to repent are the leading causes behind the environmental cataclysm that he announces to his readers.

Although John has, without question, a different conception from our own of the relationship between human actions and the corresponding "reactions" of the cosmos, there are nevertheless some clearly recognizable lines in his portrait of human sinfulness. On the one hand, this sinfulness has its roots in the fundamental failure of humankind to recognize God as Creator, "worshiping [instead] demons and idols of gold and silver and bronze and stone and wood which cannot see or hear or walk" (9:20),[31] a stance that John consistently defines in terms of "fornication."[32] On the other hand, this sinfulness expresses itself above all in "luxurious" living (18:3, 7, 9) and in violence against other human beings (16:6; 18:24). As John indicates, it is cavalier disregard of God the Creator and flagrant abuse of God's creation that result in the cataclysmic breakdown of all created things.

John has another word as well to describe this scene of devastation. What humankind has brought on itself through its failure to recognize God the Creator and to cherish God's creation is none other than the "judgment" of God.[33] It is what humankind has done to its own cosmic environment that ultimately becomes the "judgment" enacted by God against humankind. And it is nothing other than the "reaping" of the sins of the earth that fills "the great wine press of the wrath of God" (14:19–20). In Paul's words, humankind truly does "reap whatever it sows" (Gal. 6:7).

This is by no means the end of the picture. John the Revelator offers his portrait of cosmic devastation above all else in order to call people—the church in specific, but all of humankind as well—to "repentance" in the face of this present and impending disaster.[34] Matthew, for his part, shows us what such "repentance" looks like in everyday living. As Wesley Granberg-Michaelson points out, Matthew's apocalyptic discourse (chaps. 24–25) issues explicitly (25:31–46) in an "engagement in the world, feeding the hungry, giving drink to the thirsty, providing shelter to the homeless, clothing the naked, and giving comfort to the imprisoned—in short *re-establishing God's shalom within the creation* (emphasis mine)."[35] Matthew sets out this stance toward the created world and its vulnerable ones as the appropriate response to the present and impending apocalyptic devastation depicted in the previous verses. In Matthew's view this stance toward creation leads human beings to ultimate "blessedness."

CHRIST AND THE DESTINY OF CREATION

A further environmental motif within the New Testament is arguably the most crucial, namely, the destiny of creation. What will be the final disposition of that which God has created, entered into, and sustained through Christ? And what will be God's final response to the devastation of creation brought about through human sinfulness?

The New Testament writings offer a twofold, negative/positive answer to these questions.[36] On the negative side, the apocalyptic portrait painted throughout the New Testament depicts an end to the present created order. In the Gospels Jesus prophesies that "heaven and earth will pass away" (Mark 13:31). According to Hebrews, "[The earth and the heavens] will perish" (Heb. 1:11–12). And in 2 Peter the writer proclaims that "the present heavens and earth have been reserved for fire" (2 Pet. 3:7). While the descriptions of cosmic finale vary from writer to writer, the collective force of these sayings cannot be mistaken: The cosmos that we presently know and inhabit will someday come to a definitive, and apparently violent, end.

This word is nothing short of terrifying for all created beings of the present cosmic order. As finite creatures the cosmos is all that we know and all that we have. And if cosmic destruction were the final word offered by the New Testament writers, there would be no Good News to be had, no cause for hope, and no reason to concern ourselves with the question of ethics, environmental or otherwise. In such a situation the most compelling ethic would indeed be the ethic of despair identified— if not adopted—by Paul: "Let us eat and drink, for tomorrow we die" (1 Cor. 15:32; cf. Isa. 22:13).

But cosmic destruction is by no means the final New Testament word on the destiny of the created order. Beyond the warnings about destruction lie powerful promises of a very different nature. The Jesus of Matthew's gospel points forward in all confidence to "the renewal of all things," that time in which "the Son of Man [will be] seated on the throne of his glory" (Matt. 19:28). For his part Paul declares that "the whole creation," which has been "groaning in labor pains until now" (Rom. 8:22), will finally be "set free from its bondage to decay" (Rom. 8:21).[37] And John the Revelator sees "a new heaven and a new earth" in which God is "making all things new" (Rev. 21:1, 5; cf. 2 Pet. 3:13). Accordingly, if the present cosmic order is slated for destruction, this is because God is preparing a new cosmic order to take its place. As the New Testament writers proclaim it, God's original act of *creation* will ultimately give way to God's acts of *renewal and re-creation.*

Here, once again, the role of Christ becomes crucial. Christ the Agent of Creation now becomes Christ the Agent of Renewal and Re-creation. This shows up clearly in the thought world of Paul, who describes Christ as "the last Adam," "the second man," and "the man of heaven," by contrast with "the first man, Adam," "the man from the earth," and "the man of dust" (cf. 1 Cor. 15:45–49).[38] As the Genesis accounts make clear, "the first man, Adam" has been created "in the image of God" (1:27), an "image" that becomes marred when Adam disobeys the command of God and hides from God's presence (3:1–7, 8–10). In fact, Adam has a negative effect on everything with which he comes in contact. Relationships break down between Adam and God, Adam and "the woman whom [God] gave to be with [him]" (3:12), and Adam and "the ground [from which he was] taken" (3:19).

Christ reverses this all-encompassing breakdown in relationships in his role as "the last Adam." Just as "one man's trespass led to condemnation for all" (Rom. 5:18) and "by one man's disobedience the many were made sinners" (Rom. 5:19), so "one man's act of righteousness leads to justification and life for all" (Rom. 5:18) and "by one man's obedience the many will be made righteous" (Rom. 5:19). Accordingly, all of the relationships once broken because of Adam's "trespass" and "disobedience" are now restored through Christ's "act of righteousness" and his "obedience." Not only does Christ restore the relationship between God and humanity and the relationships among human beings themselves, but he also restores the relationship between humankind and the rest of the created order. As Wilkinson puts it, "this earliest understanding by the church sees the work of Christ as a 'recapitulation' of God's intention in creation. Christ is the New Adam, restoring the first Adam and his work. Just as Satan achieved power over humanity through the sin of the first Adam, he is defeated through the obedience of the Second Adam. And in that victory, Jesus reaffirms the task of the first Adam, inviting persons into renewed fellowship with God, with each other, and with the earth."[39]

Nor is this all. If the obedience of Christ "the last Adam" undoes the evil effects of disobedience by "the first man, Adam," so, too, does the resurrection of Christ undo the evil effects of the death that came about through Adam's disobedience (1 Cor. 15:21–22). Accordingly, Paul names Christ "the first fruits of those who have died" (1 Cor. 15:20; cf. v. 23) and designates Christ's resurrection as the harbinger of more to come: "But each in his own order: Christ the first fruits, then at his coming those who belong to Christ" (1 Cor. 15:23).

Here is God's ultimate answer to the devastation of creation. This answer is not death and destruction but rather "the resurrection of the body" and the "vindicat[ion of] the goodness of creation."[40] God offers this answer in the person of the Risen Christ, the living promise of future resurrection for humankind and all of the created order.

It is this portrait of Christ and his role in God's "plan for the fullness of time" (Eph. 1:10) that we see in the Deuteropauline Epistles. In Eph. 1:9–10 we learn that God "set forth in Christ . . . a plan for the fullness of time, to gather up all things in [Christ], things in heaven and things on earth." The letter to the Colossians describes this "gathering up" action: "For in [Christ] all the fullness of God was pleased to dwell; and through him God was pleased to reconcile to himself all things, whether on earth or in heaven, by making peace through the blood of his cross" (Col. 1:19–20).

In short, it is the entire cosmos, and nothing less, which is the object and the recipient of God's saving action carried out through the death of Christ.[41] God's redemption through Christ is the redemption not only of humankind but of the entire created order. As Wilkinson comments, "The Gospel [is] good news for all of creation, not just humans . . . Redemption is not human salvation *out of* a doomed creation, but rather the *restoration* of all God's purposes in creation."[42]

The implications of this cosmic "good news" for our lifestyle and actions are fundamental and far-reaching. If God in Christ has paid the ultimate price in order to "reconcile [the cosmos] to himself" (Col. 1:20), then we as disciples of Jesus Christ are called to align ourselves as well on the side of the cosmos and to engage our efforts boldly on behalf of "God's purposes in creation." And if God's response to cosmic evil has been the resurrection of Jesus Christ to new life beyond death, then we as heirs to that resurrection life are called to "be steadfast, immovable, always excelling in the work of the Lord, because [we] know that in the Lord [our] labor is not in vain" (1 Cor. 15:58).

CHAPTER 9

Pacifism, Nonviolence, and
the Peaceful Reign of God

Walter Klaassen

In October 1970, my family and I found ourselves at the end of a gravel road in the upper reaches of the Unterberg Valley in the Tyrolian Alps. It was heavenly, the beauty indescribable—the clear, crisp autumn air; the free, cold brook; the mountain silence. But sin had entered this Garden of Eden not long before, for beside the brook—indeed, in it—were the remnants of someone's fast-food picnic, complete with its plastic carrying bag. It seemed a desecration; we gathered up the rubbish and carried it back down.

A year or two later we were motoring in northern Ontario. On a beautiful sunny day about lunchtime, we were looking for a suitable picnic site. We found one, a river rushing deep and cold down to Lake Superior. But when we got out of the car our expectations evaporated in a stench of acid. Then we noticed that the river was dark brown and that the vegetation that had once graced its banks was the gray color of ashes for two or three meters on either side. The river had become a waste disposal conduit for a pulp and paper mill upstream. Sick at heart, we drove on.

In November 1978 we had escaped from the crowding and noise of Rome and decided to take a few quiet hours at a beach. We followed a sign directing off the autostrada and quickly found ourselves on a completely abandoned beach on the shore of the Adriatic. Actually, there was one lone man there, fishing. We jumped out of our camper; the children ran along the sand, enjoying their freedom from the restrictions of the last few days; Ruth and I stood still to enjoy the beauty of the quiet seascape. When we returned to the car to continue on to Pisa, we brushed the sand off our feet, or tried to. We found to our horror that we had sticky oil on our feet and had to siphon some gasoline from the car with which to wash the oil away. We went back to the beach and found little sand-grain-sized globules of oil everywhere.

THE VIOLENCE AGAINST MOTHER NATURE

These three experiences, among others, shocked me into the awareness of real and sinister threats to the planet. It may be that my dismayed response had been conditioned in the late 1940s by the reading of two books by Ray Stannard Baker. Baker, under the pseudonym of David Grayson, wrote on living with exultation in the natural world of soil, sun, woods, and the ordinary tasks that made up his life as the farmer of a small holding somewhere in New England. I read his books at most two years after I had left farming for good; they taught me many things that I already knew and also much that I had missed when I was on the land. I was impressed by Baker's love for the natural surroundings into which he had fled, his lack of aggressiveness, and the dominance of a spirit of cooperation with the living environment around him.[1]

It is over twenty years now since I began to think of the response to human violence against our living home in terms of the Mennonite tradition of nonresistance or, better and more literally, living without weapons.[2] In the decades since World War II, we had done some careful thinking about pacifism, nonviolence, and living without weapons in the social and political culture of our time. However, as far as I am aware, we had done *no* thinking about the resources of our tradition of nonviolence in the human war against mother nature.

We know well how difficult it is to articulate persuasively to our contemporaries how one lives without weapons in the modern world. We are up against an ancient human legacy, succinct and efficient—our penchant for forceful and violent response to injury, be that in domestic, social, or international relations. Coercion, threat, and killing are effective in solving social problems in the short run. The acceptability of this approach seems, on the whole, to be unquestioned, buttressed by appeal to private right, patriotism and nationalism, rights of possession, and ancient fear and prejudice. As a society we give lip service to the excellence of nonretaliation and forgiveness but quickly insist that, in the harsh realities of life, when it really matters, a nonviolent response is not practical. After all, even Mahatma Ghandi failed to solve India's problems with his nonviolence.

Powerful positions clash today with the more recent recognition that we need to adopt a nonviolent, living-without-weapons response toward the natural environment of which we human mortals are a part. What are these influential orthodoxies that still go largely unquestioned? An obvious one is a passionate belief in the absolute right to private possession. The ancient Greek myth of the Golden Age identified greed as the original human sin, which quickly expressed itself in claims to private ownership. The encounter of European acquisitiveness with the aboriginal people of North America has often been described. Canadian novelist Rudy Wiebe wrote an especially moving passage in *The Temptations of Big Bear*, in which the venerable Chief Big Bear asks of the conquerors this question about private ownership: "Who can receive the land? From whom would he receive it?"[3]

Closely linked to this belief in the absolute right of possession is the conviction of the unimpeded right to pursue wealth in Western industrial culture, which has translated into regarding the earth with its treasure as existing solely for the enrichment and power of those who can take and into regarding what we call *natural resources* as located in the current rather than in the capital account of nature. A recent adaptation of this view is the conviction of the right to abstract and force natural processes into technical means for the power and enrichment of the already powerful and rich. One thinks of genetic engineering to control plant and animal production, to eliminate open strains and variety in the gene pool, and then patenting of the end product for absolute control. All of this has been made possible in large measure by a trick of the mind devised by Western philosophy in which human beings are set over against the world in which they live, making them the detached, subjective observers of objective nature and then taking a further step away in denying human kinship with the rest of creation. Thus, we are made human spectators, indeed aliens, in our own home.

Who knows how all of this happened? That these orthodoxies are firmly in place in our culture no one would care to dispute. It took a long time for them to develop. The fiction of the right to private possession comes at least with the rise of the cities in the cradles of the ancient civilization. The fault is not simply in the Judeo-Christian tradition, as is still popularly claimed. These ancient orthodoxies found fertile ground as the West rose to world power beginning with the development of banking in the Italian city-states and with the emergence of the commercial empires in the seventeenth century. It is a truism that religious people, in this case Christians, come to reflect the culture they helped to create.

ANABAPTISM'S VIOLENCE AGAINST NATURE

Unfortunately, the Radical Reformation tradition, which could have been a defense against these old orthodoxies, was not (with one exception, to which I shall return later). Anabaptism was initially an urban movement, and a mere century after its beginnings, urban Mennonite merchants in Amsterdam, still committed to nonviolence, were already arming their trading ships with cannons to deter pirates or privateers. Elsewhere in Europe the Mennonite adoption of agriculture as an exclusive way of life seems not to have been intentional. That Mennonites came to outrank others in the craft of agriculture was not necessarily because they were more devoted to caring for the creation, in particular the land, but because they were a small, persecuted group, struggling to survive. Their agricultural skills were developed to that end. And because agriculture was all that they did, all their inventiveness was lavished on it. It was subsistence agriculture, and they used every trick of the trade known or discovered by them to ensure survival and the goodwill of the landlords whom they served. By the time they became independent farmers, they had acquired the technical advantage that in large measure explains their success in Prussia, Alsace, Russia, and North and South America, even under difficult conditions.

It was the need to survive and not love of the land that produced the expertise and care of the land for which Mennonites became famous.[4] I make this judgment because nowhere in the literature on Mennonite

agriculture do I find that their care of the land was motivated by any biblical imperative to care for the creation or that to love God means to love
the land. The tradition of living without weapons in a dangerous human
world was consciously derived from Scripture: "He who loves God must
love his brother also." Not so the care of the land.[5] I say this reluctantly
and may be proved wrong, but I sense among our people little love of the
land, little appreciation of the beauty of nature as related to the vocation
of agriculture.

We did not have a clear understanding of a God-given mandate to
farm without weapons as we had a mandate to live without weapons with
our fellow citizens. Further, we became expert tillers of the soil in order
to survive. Because of these factors we developed or accepted without
question all of the most recent techniques for increased production. Consequently, we have not had any defenses against the industrialization of
agriculture, with its threat to the productivity of the land. Some, like the
Amish, have resisted the industrialization of agriculture, but with decreasing success, and the Hutterites have bought into agricultural technology with little resistance. All of us in the Anabaptist family have become willing and often enthusiastic contributors to technological hubris
and its ally consumerism, which, in the words of a recent writer, "has
achieved a . . . moral . . . sovereignty."[6]

That there was here and there a Mennonite Aldo Leopold or a John
Muir is readily granted. That we have had a reputation for thrift and
honesty and moderation, likewise. But we need to abjure any romanti-
cism suggesting that somehow care for the land was part of our ethnic-
ity or that we have been better than others in caring for the earth; if we
ever had it, it is certainly gone now. *The Farm Report* comes out of the
Land Institute near Salina, Kansas, and is devoted to agriculture that
cares for and nurtures the land. Why do I not encounter the name of a
single one of the hundreds of Kansas Mennonite farmers whose ancestors settled there in 1874? Today's consciousness among Mennonites
about the need to care for the creation has not come from our own tradition. We have listened to others in the church and out of it, and we are
beginning to catch up. We, too, are beginning to understand that "the
peaceful reign of God" is not limited to God's human children but includes the whole creation.

THE PEACEFUL REIGN OF GOD

In what I am about to say, I am taking a deliberately narrow approach to the subject, leaving out of consideration many larger issues such as "the peaceful reign of God" vis-à-vis national states and the whole political realm. I am deliberately taking a personalized approach to "the peaceful reign of God" because this approach has too often gone begging in our preoccupation with larger social issues.

I don't need to detail again all the components of the gathering darkness of ecological destruction. We know them only too well, and new details are added every morning. So, do we despair by giving up and, perhaps reluctantly and even with sorrow, continue to eat, drink, and be merry and "after us the deluge?" Among the heirs of the Radical Reformation, there are many who have chosen this response. That is what comes to mind when I see the enormously large and rich Mennonite churches in the Fraser Valley of British Columbia and learn of their decreasing contribution to the relief of the world's suffering through the Mennonite Central Committee. Or do we invite despair by exhausting ourselves in special issue projects that tend to increase our frustration and anger and resentment at powers we cannot control or turn to our ends? No, as Christ's followers and as God-appointed caretakers of the creation, we do not. Rather, we "give thanks to God, who gives us the victory through our Lord Jesus Christ. Therefore, beloved sisters and brothers, be steadfast, immovable, always excelling in the work of the Lord, because you know that in the Lord your labour is not in vain" (1 Cor. 15:57–8).

Christians in Europe and especially in North America in the last 150 years quite misunderstood the nature of "the peaceful reign of God." By imperial political, social, and missionary strategies, it was believed, it was possible to bring about "the peaceful reign of God" here and now in our human space and time. Everywhere Christians saw signs of the coming of God's kingdom. The hymns were wonderful.

> *For the darkness shall turn to dawning,*
> *And the dawning to noon-day bright,*
> *And Christ's great kingdom shall come to earth,*
> *The kingdom of love and light.*[7]

Or the following:

> *Hail to the brightness of Zion's glad morning,*
> *Joy to the lands that in darkness have lain!*
> *Hushed be the accents of sorrow and mourning,*
> *Zion in triumph begins her mild reign.*
> *See from all lands, from the isles of the ocean*
> *Praise to Jehovah ascending on high;*
> *Fallen the engines of war and commotion,*
> *Shouts of salvation are rending the sky![8]*

It would be churlish indeed to complain about words like that; in fact, I love those hymns, for whenever I read or sing them, my longing for the vision they express increases. The problem is that we confused liberty, equality, and fraternity (the "rights of man"), life, liberty, and the pursuit of happiness, and Westernization and democratization achieved by imperial political means with the coming of "the peaceful reign of God." And when the already visible gleaming spires of the city of God retreated like a mirage, blocked from sight by the incredible iniquity of the great wars and the cold war and the ever-increasing hunger and violence and environmental destruction, we North American Christians wore ourselves out in frustration and disappointment. Knowing that our own motives were genuine and honest, we located the sources of the problems "out there" in governments and institutions and perverse people when we sang "Oh, when will they ever learn?"

Mennonites joined the action rush, and we redoubled our efforts to compel into existence "the peaceful reign of God" through the many programs of the Mennonite Central Committee, Mennonite Disaster Service, Voluntary Service, Christian Peacemaker Teams, and many others. These programs are honored and they are important, good, and worthy of our support. But the peculiar temptation that comes with organizations and programs is to locate the evil we oppose, the resistance to and the frustration of "the peaceful reign of God," out there and not in here.

"Those conflicts and disputes among you," we read in the Christian Scriptures, "where do they come from? Do they not come from your cravings that are at war within you? You want something and do not have it; so you commit murder. And you covet something and you cannot obtain it; so you engage in disputes and conflicts" (James 4:1–2).

THE INDIVIDUAL'S ROLE IN THE KINGDOM OF GOD

We continue to talk about building the kingdom of God as though *we* have to establish it, as though it is *our* task. After all we have experienced and learned this century, we still think that God expects us to do it and that we can! We believe desperately that somehow it will clearly and without spot or wrinkle succeed the present witches' brew of greed, oppression, violence, and domineering ambition, of smallness and narrowness of heart, and of deliberate, invincible ignorance.

The great church father Augustine was far closer to the truth when, looking out into the darkening world of the early fifth century, he saw a society and a church, both of which were a mixture of good and evil, and that the coexistence of good and evil in human life in all its manifestations was a permanent condition. I learned what I used to think was an excessively pessimistic view of things from my own experience in the peace movement and in a three-year stint as co-chair with Ruth, my wife, of the Environment Working Unit of the British Columbia Conference of the United Church of Canada. Many of the people in the peace and environmental movements, I learned, were just as vengeful, impatient, intolerant, moralistic, unjust, contentious, unpeaceful, and, indeed, violent as the people they opposed. Just recently, for example, a group that calls itself "environmental" made death threats against government and industrial leaders in British Columbia.

I learned to my sorrow and near despair that this proneness to vengeance, moralism, and violence was in me, too. In these movements we profess to be different and seem often naively unaware of our hypocrisy. We are very reluctant to acknowledge that the evil that spawns the violence in Rwanda, Bosnia, and Haiti is also in us. We Mennonites should know this by now because recently that evil and violence has been repeatedly and publicly shown to exist in our very homes and churches.

I no longer believe that there will ever be a human society on this earth from which evil and injustice will be absent. The beautiful visions of Isaiah 11, Romans 8, and Revelation 21 and 22 of peace and plenty in nature, human society, and the cosmos are just that, visions, images of a healed and restored humanity and creation. They are bright, celestial beacons that illuminate the harbor entrance. They are like the large, red, flashing lights on the hilltops of the Okanagan Valley where I live, which provide a corridor along which aircraft can fly safely in the darkness to the airport at Kelowna. These visions are the indicators of potential; there is that within us that can see and hear and that can choose to take up direction toward those visions.

The seeing and the desire and the will to go toward those visions is an intimation of the presence of "the peaceful reign of God." For "the peaceful reign of God" is here and now; it was given by God, revealed to us in the supernova appearance of Jesus in our time and place in the incarnation, God's self-revelation in flesh and blood to us "upon whom the ends of the ages have come."

The incarnation, too, has to do not only with God's human children. It is this great, unique Christian assertion that best symbolizes the relationship of God to the creation. Incarnation means "enfleshment"; God took matter and human personality as a medium of revelation. God pronounced the whole material creation "very good" at the beginning; in the incarnation God, as it were, underscored in red that original goodness. Whenever, therefore, we see a stone, a tree, an animal, or a person, by faith we see the speech of God. They are signs of the reality and presence of "the peaceful reign of God" (1 Cor. 10:11). The incarnation is the beginning of the end-time, "the times of refreshing . . . from the presence of the Lord, . . . the time of universal restoration that God announced long ago" (Acts 2:20–21), the restoring of all things to their primal splendor. The end-times are now.

THE INCARNATION AND THE PEACEFUL KINGDOM

"The peaceful reign of God" began for Christians with the incarnation, the "dawning upon the world of healing for all humankind" (Titus 2:11). There is no talk anywhere in the Christian Scriptures of remaking the political or social systems of our time; there is a great deal about learning to recognize and welcome "the peaceful reign of God," about repentance and lining up with God's intentions, grasping the reality of what has already been given. "Do not be afraid, little flock, for it is your Father's good pleasure to give you the kingdom" (Luke 12:32).

There is no need to strain for it, much less to build it; the edifice is already there. What we need desperately are the eyes of spiritual discernment to see it, ears that can sort out what God is saying to us from all the other voices that clamor for our attention, and hearts that are able to filter out and embrace the truth from the welter of lies and falsehood around us. It is for us a matter of nonresistance, of ending our resistance, to the scriptural insistence on the new reality of the already given and established "peaceful reign of God." We North Americans especially resist the conclusion that this reality of "the peaceful reign of God" is already given and that all our efforts and technological expertise, all our grunting and pushing and shoving won't make it any more real than it already is. *Gelassenheit*, some of our Anabaptist ancestors called it. "Letting go and letting God" is how it is more popularly stated today.

This letting go of the idea that we build the kingdom, that we must in our generation bring it to its completion, this surrender of our conceit that it is all in our hands and dependent upon us, this is the true nonresistance, the primal, foundational, and requisite nonviolence, the surrendering of all the weapons of human independence, pride, control, and domination. Here is where it begins; it becomes a personal search and struggle and victory for each person. The problem has first to be dealt with in the internal life of each of us. It is learning what James meant when he wrote to Christians: "You want what you cannot have, so you murder; you are envious, and cannot attain your ambition, so you quarrel and fight." It is learning that so much of our anger and frustration comes from the tyranny of elevating what we consider to be urgent over what the Scriptures tell us is important.[9]

The visions of "the peaceful reign of God" in Isaiah 11, Romans 8, and Revelation 21 and 22 offer a lot of specific details: peace within the animal kingdom, especially between wild and domesticated animals, and the total absence of injury and destruction; the liberation of the creation from entropy; the New Jerusalem from which all destruction and chaos have been expelled, where human pain and evil no longer afflict and no evil threatens, so that the gates are permanently open and the communion between the eternal God and the creation will be completely restored. Those are not predictions of a literal, universal state of affairs within human history. They represent a different angle of vision, a hope, a yearning for what may be, an intense desire for the consummation of "the peaceful reign." In another sense, however, we need to take them literally.

When we make them our own and internalize them by sharing the hope for their fulfillment, we yield to their potential power. We live in God's kingdom now; we are its citizens now. We ourselves participate in fulfilling the hope and the vision by seeing and hearing and doing like the citizens of God's kingdom now. So we do not hurt or destroy in God's holy mountain, the whole creation. We live without weapons and nonviolently each day in the midst of the creation, loving and protecting the creatures against injury, making their life and growth possible, and regarding all created things as kin because we share the same breath of life with them.

We take it upon ourselves to satisfy the eager expectation of the created universe for the revealing of God's sons and daughters. Indeed, we are the active promoters and supporters also of public policies to preserve the forests, the waterways, the air, the animals, birds, and fish in danger of extinction, enhancing the life of all. We are not private individuals; we are part of the whole.

THE PEACEFUL REIGN: BEGINNING
WITH OURSELVES

I do not want to be understood to decry or devalue public action for the betterment of humanity and the creation. I have given a lot of energy and time to those actions in my life and continue to do so. But I am coming in by what for many liberal Christians has become a back door. Rather than beginning with the bad people and the evil out there in society, I want to begin with myself. This is a moral and not a temporal order of precedence. What we publicly urge and do can help to give visibility and concreteness to "the peaceful reign of God" also when it is done by governments and institutions. But we act first, not necessarily in the sequence of time but in the order of moral priority, like the revealed children of God where we are. And that is *not* a cop-out. It is far more difficult than appears at first glance because what is required is a radical change of habits, an awareness that even where I am at home I always come up against the public orthodoxies that tell me that all of this is not only futile but also ethically questionable because it will change our whole way of living based on the never-ending cycle of production and consumption. The jobs of our fellow citizens are at stake! For the most part, no one compels us to do these things; only our new seeing and hearing will tell us that it is right because we are citizens of "the peaceful kingdom."

Unless we can begin to act as the revealed children of God at home, we will have no resources of soul for public action guided by the biblical vision. In fact, our tendency to concentrate on programs of public action may distract us or excuse us from personal responsibility for the creation where we are. Only with those inner resources will we have staying power for the long haul of public action. The willingness to surrender nonresistingly to the present reality of God's peaceful reign will allow us to be satisfied to have God and the Lamb to be our light; this divine light will somehow enable the nations to find their way without weapons of aggression and control because the gates of the city will always be open, with no cannons threatening those who approach it. If we internalize this vision and make it our own, if we can take up residence in the vision of John of Patmos, we will ourselves become the channels for the water of life flowing from beneath God's throne, welling up within us and flowing out from us to water the trees of life, whose leaves are there for the healing of the nations.

We have the privilege of becoming all this, not by creating consumer-oriented world spirituality empires and high-priced ashrams and opulent ego-stroking retreat centers, but by being God's own caretakers in the small places where we live with as much of the creaturely world around us as those places allow. The chickadees and the fir trees, the coyotes and the black bears won't tell anyone about it; they won't appear on cable and network TV talk shows to tell the world what great people Ruth and Walter Klaassen are; no one will know about most of it, but such living is expressing an awareness of and giving visibility to "the peaceful reign of God." And the creatures will "know" that their Creator's human sons and daughters are being revealed.

I recently found a poem by David Gordon Helwig, who teaches at Queen's University in Kingston, Ontario. This modern poem is about the land in which I now live. It bears witness to the evocative power of the vision of Isaiah and perhaps also of the famous paintings of the Peaceable Kingdom by Edward Hicks. It startles by its use of the names of animals that live around us where we now are. This poem bridges the gap of twenty-seven hundred years and makes the vision contemporary. The title is "One Step from an Old Dance":

> *Will the weasel lie down with the snowshoe hare*
> *In the calm and peaceable kingdom?*
> *Will the wolverine cease to rend and tear*
> *In the calm and peaceable kingdom?*
> *Will the beasts of burden not have to bear?*
> *Will the children feed grass to the grizzly bear*
> *In the calm and peaceable kingdom?*
> *Oh the wolverine will cease to tear*
> *In the calm and peaceable kingdom,*
> *The rattlesnake rattle praise and prayer*
> *In the calm and peaceable kingdom.*
> *Oh the wolves will wear smiles like the children wear,*
> *The wolverine will cease to tear*
> *While the hawk and the squirrel are dancing there*
> *In the calm and peaceable kingdom.*[10]

None of this is concerned with programs and organizations. Our actions on behalf of the creatures are not important in that they become an irreversible force that will win in the end, so long as we doggedly keep at it. These actions are ends in themselves; they are intimations of "the peaceful reign of God" and need no further justification.

THE PEACEFUL REIGN: BOTH PERSONAL AND CORPORATE

What I have attempted to describe is not a privatizing of Christian faith and life, but a personalizing of it. It is not spiritualizing the historical and corporate nature of Christianity. The personal and the corporate, the historical nature of Christian faith and individual piety, should never be severed. But whatever dimension our faith exhibits, it always truly begins with the discovery of the pearl of great price, finding that one is loved by God and seeing the beauty and greatness of God. It may also eventually blossom into the surprising recognition of the mystery that God "determined beforehand in Christ, for him to act upon when the times had run their course: that he would bring everything together under Christ, as head, everything in the heavens and everything on earth" (Eph. 1:9–10 NJB). The personal is by definition part of the greater whole of the human community and, for Mennonites especially, part of the church. The community of the church is where we hear the gospel; it is the laboratory in which we learn to trust and love and where we become aware that we are part of a movement across time and space, called and dedicated to the ministry of reconciliation, which now means the reconciling not only of human enemies, but equally of the human and nonhuman creation.

I love to return again to the Scriptures, to those visions seen by the prophets and apostles and singers of Israel about the "peaceful reign of God." There is a strange concatenation of judgment and celebration in some of the Psalms, especially Psalms 96 to 99 and 104. Psalm 99 begins: "The Lord is king; let the people tremble!" In many Christian circles it is today politically incorrect to speak about God as king, as reigning, as judging, and instead God is portrayed as a morally nondiscriminating, indulgent Santa. Such an attitude represents the deliberate denial of a theme that runs through the Bible from beginning to end. "The Lord is king," and one of the functions of a king was to be a judge, to dispense justice.

"He is coming to judge the earth; he will judge the world with righteousness, and the peoples with his truth" (Ps. 96:13). How may we, today, in such a different time and place, understand these words? Indeed, "let the peoples tremble!" For God comes to us here in America with his truth to lay bare the terrible travesty we have made of human nature, so much so that it has become a commonplace in our consciousness and in public discourse. Human beings have been degraded from being created in the image of God, with all the richness and potential that implies, into consumers. God comes to judge us for this insult to the creation, a blasphemy that most Christians accept with barely a murmur of dissent.

God comes to judge the disruptions and the extinctions we have caused and are causing in "the peaceable kingdom." We gradually crowd the creatures from their natural habitat and make it inhospitable, all to feed our voracious appetite for consumer goods. If we have such a vulgar view of ourselves, how can we have any respect for the rest of the creation? That judgment is in the first instance the judgment of the gospel in our own lives. It is the judgment upon our treatment of God's creation for the satisfaction of our appetites, for being content and comfortable in our role as "consumers." "The least of these," to whom the Judge in the Parable of the Last Judgment alludes, must now also include all our fellow creatures on this planet. If we repent, we will allow the light that has come into the world to judge and dispel that terrible darkness of distortion in each of us. When that darkness begins to lift, the children of God are revealed. This is what the composer of the 104th Psalm meant when he said: "Let sinners be consumed from the earth, and let the wicked be no more" (Ps. 104:35).

The judgment that makes the people tremble is the occasion for the joy of creation. Listen to the music of it:

> *Let the heavens rejoice and the earth be glad!*
> *Let the sea thunder and all it holds!*
> *Let the countryside exult, and all that is in it,*
> *and all the trees of the forest cry out for joy,*
> *at Yahweh's approach, for he is coming,*
> *coming to judge the earth;*
> *he will judge the world with saving justice,*
> *and the nations with constancy. (Ps. 96:11–13 NJB)*

That is the judgment that makes the people tremble and makes the creation rejoice. When that cleansing judgment takes place in me and in you, we will begin to understand and, in our turn, to rejoice in "the peaceful reign of God" and our own place in it. The 104th Psalm expresses, as does no other passage of Scripture, the relative position of human beings in the creation. This Psalm, with its thirty-five verses, directly alludes to human beings only once. It is like the landscape painting of the sixteenth-century Chinese painter Wen Chen-ming, in which human beings are disappearingly small. In this Psalm human beings are seen as part of nature and not as its lords. We are co-creatures with animals and trees, water and air, and cannot exist independently. If this understanding has not been part of our Anabaptist heritage from the beginning, we have the opportunity to make it part of our tradition and part of the tradition of Christian faith now, in our own time.

CHAPTER 10

An Anabaptist/Mennonite Theology of Creation

Thomas Finger

Over the centuries, most Anabaptist/Mennonites have been agricultural peoples whose existence has been closely intertwined with the natural world. Surprisingly, however, almost no theological reflection on the doctrine most closely associated with nature, that of creation, has arisen from these groups. The original Anabaptists said little on this theme.[1] Most Mennonite confessions of faith either mention creation only briefly or omit it altogether.[2] Why is this? One reason is surely that Anabaptist/Mennonites have been intensely practical rather than theoretical people and have shown little inclination to reflect intellectually on most formal theological topics.

Yet there is a deeper cause. The early Anabaptists viewed their world dualistically, as torn by conflict between God's kingdom and the kingdoms of this world.[3] Some apparently understood this dualism in an "ontological" manner, which linked the natural creation with the fallen institutions and people around them, and they expected all this to be destroyed at Jesus' imminent return. Others interpreted birth into God's kingdom in so spiritual a fashion that they apparently expected the physical world to pass away entirely at the End.[4] In both cases, the nonhuman creation would have no lasting value and would occasion little concern. Does this mean that an Anabaptist/Mennonite perspective must ignore, minimize, or downgrade the natural creation and can provide no orientation for environmental issues?

In probably the only article on a Mennonite theology of creation published in recent years, Calvin Redekop points the way beyond such a conclusion. The early Anabaptists, he maintains, were so overwhelmed by their life-threatening struggle in the social and ecclesiastical realms that they confined their discussions of good and evil to these spheres. Though some may have assumed an ontological dualism, none really had occasion to think seriously about nature's role in all of this. Consequently, early Anabaptist theology was incomplete, yet open to innovative reflection on creation should the need for it later arise.[5] Moreover, Anabaptist experience, both early and continuing, did contain themes that pointed toward fuller consideration of creation.

Early Anabaptists expected God's kingdom to be lived out as fully as possible within the Christian community, no matter how soon or far off was Jesus' return. Such a lifestyle involved all the community's activities, including labor with and use of natural creatures. But such living is impossible without a supporting nonhuman environment, impossible unless the nonhuman creation can also be indwelt and transformed by God's reign.

Anabaptist dualism, Redekop seems to say, is not really "ontological" (descriptive of its essential reality), not between realms of matter and spirit. Instead, it focused on ethical and religious dualism—between choices and life directions that humans adopt.[6] But this latter by no means entails a negative view of nature. Redekop maintains that early Anabaptism implicitly pointed toward "the full-orbed concept of shalom: the Hebraic notion of peace that includes transformation of the physical creation."[7]

Continuing experience revealed more fully how creation could play a role in God's kingdom. During their first fifty to sixty years, skillful care of the land enabled Anabaptists to survive and win toleration, and they developed a reverence for the land. Over the next several centuries, such agricultural activity provided the basis for the erection of stable Mennonite communities and institutions.[8] In these ways, later experience filled out and made available for theological reflection a dimension of faith that earlier Anabaptists had overlooked.

But if an Anabaptist/Mennonite theology of creation is possible, from what vantage point should one embark upon it? Redekop suggests expanding on some ethical emphases central to Anabaptist kingdom activity. For instance, one could show that a consistent nonviolent stance forbids the exploitation of nature.[9] However, since no real theology of creation in Mennonite perspective has been formulated, other starting points consistent with Mennonite theology and praxis can be proposed. Having shown, by following Redekop, that Anabaptist/Mennonite awareness of the presence of God's kingdom entails a positive role for nature in God's salvific activity, let me suggest another way of beginning from the center of the Anabaptist/Mennonite perspective.

WHERE SHALL ANABAPTIST/MENNONITE THEOLOGIZING BEGIN?

What is the focal point of Anabaptist/Mennonite commitment, praxis, and theological understanding? I propose that it is best expressed not simply as the kingdom of God, nor as community, nor as peace, but as the process that brings everything under the radical, living lordship of Jesus Christ. This means, broadly speaking, that theological reflection should begin with redemption through Jesus. Whereas traditional confessions and systematic theologies start with creation and move toward redemption, a distinctive Anabaptist/Mennonite theology should first consider redemption—or the new creation—and then in that light discuss the original creation.

Biblical scholars generally agree that this was the route followed in the Old Testament. Israel's experience of redemption from Egypt provided the perspective from which Israel came to believe in and speak about God as Creator.[10] Similarly, the recent Mennonite Confession of Faith, though following the traditional order by placing creation articles before articles about redemption, includes the coming of the new creation in Christ and the Spirit and the anticipation of creation's final redemption in its summary statement about creation.[11]

Anabaptist thought, which has generally begun from the redeeming work of Jesus, has insisted that this cannot be narrowed to his death and resurrection but must also include his life and teachings and the necessity of our following them. During the last fifty years, this insistence has taken the form of summarizing an Anabaptist vision in terms of "Mennonite distinctives" derived largely from Jesus' life and teachings: discipleship, brotherhood, and peace/nonresistance.[12] Although such emphases are essential to Jesus' redeeming work, it is possible to interpret them simply as ethical ideals and to reduce Jesus to someone who once promulgated them and exemplified their truth.

If this route is followed, an ethical ideal can become the focal point and norm of Mennonite theologizing instead of the active process that brings everything under the radical, living lordship of Jesus Christ. This could result—though it would not need to—from building a theology of creation on a theme like nonviolence. Moreover, the life and teachings of Jesus alone cannot provide a sufficient basis for a full-orbed theology of creation. In fact, environmental issues may well be showing Mennonites the inadequacy of reducing the significance of Jesus Christ to this dimension. Nonetheless, since his life and teachings are essential aspects of this work, let us begin by ascertaining what they can tell us about creation.

If we construct a theology of creation beginning with Jesus' life and teachings, we will be impressed with profound reverence for creation's intricacy and beauty. This reverence shines through many of Jesus' parables and sayings (esp. Matt. 6:28–29) and echoes the frequent expressions of wonder at and thankfulness for creation found in the Psalms, Isaiah, and elsewhere in the Old Testament. Jesus also taught that nature's regular rhythms express the universality of God's rule (Matt. 6:45), another foundational Old Testament theme.

Moreover, Jesus emphasized that God is especially concerned for the weakest and apparently least conspicuous creatures—for the grass that quickly withers and is burned (Matt. 6:28–30) and for sparrows, about the only meat source available to Palestinian peasants.[13] This concern for the weakest nonhuman creatures paralleled Jesus' special attention to the most unfortunate and victimized human creatures. It was consistent with Yahweh's historic favor toward the inconspicuous, the needy, and the oppressed.

Significant as these themes from the life and teachings of Jesus are, they do not enable us to answer some broader questions about creation that arise in light of contemporary environmental issues. From some conservative Christians, one hears that, since God will soon destroy creation (echoing some early Anabaptists), we need not worry about its preservation or betterment. In response one can, like Redekop, argue that Jesus envisioned a *shalom* that involved some continuance of nature. Yet many other scholars have interpreted Jesus as an apocalypticist who envisioned nature's imminent destruction, and some sayings of his point in this direction (esp. Luke 21:9–28 and parallels). Moreover, other Old and New Testament texts seem to express a similar vision (e.g., 2 Peter 3:10–12). Adequate response to this "conservative" claim seems to require broader theological considerations.

From the more "liberal" side of current environmental discussion, however, one finds Christian theologians making quite different affirmations. Environmentally concerned people stress the importance of recognizing that all earthly creatures are interconnected. In contrast to traditional Christian theology, these theologians argue that interconnectedness will best be appreciated if we regard the being of God not as distinct from his creation, but as intertwined with the being of creatures. This view is called *panentheism*, which means that God (*theos*) is in (*en*), or inseparably interconnected with, all things (*pan*).

Sallie McFague and some Process theologians recommend envisioning the world as God's "body."[14] If we regard each creature as part of God, they argue, we will be likely to reverence it and work for its benefit. But if we regard its being as distinct from God's, we will probably suppose that God is distant from it and have far less concern for its wellbeing. Process theologians, Rosemary Ruether, and other writers also argue that appropriate Christian environmental attitudes can be inferred from the evolutionary process that produced our present world.[15] This estimate of evolution differs considerably from the more traditional one, according to which evolutionary struggle licensed the kind of combat with nature that spawned many current environmental problems.

Is God intertwined with nature or about to destroy it? Can evolution provide the basis for environmental praxis? Such questions raise cosmic issues about God's relation to the universe, which cannot be fully answered by an ethically oriented approach based on Jesus' life and teachings. In other words, the environmental challenge pushes Anabaptist/Mennonites to ask whether another way of considering the redemption brought by Jesus—of considering how everything is coming under his radical, living lordship—might provide an avenue toward answering cosmic theological questions.

THE WAY TO THE CROSS

Anabaptist/Mennonites have not only emphasized the life of Jesus. They have also stressed that his way of life led consistently toward his death. It is possible, then, to consider Jesus' life and teachings under the broader theme of his way to the cross. In so doing, however, it is important to avoid several oversimplified dichotomies.

The first is to suppose that some distinct "Jesus of history" can be separated from a more cosmic "Christ of faith." The overall structure of each gospel indicates that it intends to show that Jesus was much more than a merely human, historical figure. (This can be illustrated easily by showing how a thematic passage from the beginning of each gospel correlates with a climactic one at the end: for example, Mark 1:1 with 15:39, Matthew 1:23 with 28:19, Luke 1:35 with 24:46–47, John 1:1 with 20:28.) A second dichotomy is to suppose that Jesus' cross and his way to it can be separated from his resurrection. Yet all the gospels climax with the resurrection and point the way toward it from the beginning.[16]

Moreover, it is ahistorical to suppose that early Anabaptists were not really concerned about the resurrection of Jesus or cosmic significance. Such claims may seem plausible if one assumes that the "distinctives" they stressed (involving his life and teaching) could be separated from their overall view of Christ. But no Jesus reduced to these dimensions existed as an object of faith in Reformation times. Moreover, Alvin Beachy has convincingly argued that Anabaptism held a very distinctive view of salvation: It was "divinization," or renewal through participation in the energies of Christ's divine nature.[17] For Christ to play this distinctive role, he clearly had to be resurrected and cosmically active. Indeed, his resurrected activity provided the energy for living the ethical life that Anabaptists stressed.

Starting with Jesus as presented in the gospels, let us see whether a consideration of his way to the cross, which Anabaptist/Mennonites find essential to his redemptive activity, might eventually yield some insights concerning creation. I begin with a phenomenon seldom stressed by Mennonites, yet one deeply ingrained in the gospel records: The activity of Jesus was inseparably intertwined with the actions of his Father and Spirit. The way to the cross for Jesus cannot be adequately understood apart from the commissioning (Mark 1:11 and parallels), command, and continuing love of his Father,[18] nor apart from the commissioning (Mark 1:10 and parallels), guidance, and empowering of the Spirit. Jesus did not accomplish what he did on his own, but only through cooperation with these two other agencies.

Similarly, Jesus' cross involved not only himself, even though he died with a horrible sense of abandonment (Mark 15:34). Early Christians perceived the Father not as distant from the cross, but as giving up his Son through a profound act of costly love (Rom. 5:8, 8:32; 1 John 4:9–10). As Juergen Moltmann has said, the Son experienced the pain of actually dying, but the Father experienced the grief of one whose beloved dies. In this dark experience of separation, the two were united by the Spirit, through whom Jesus offered up himself (Heb. 9:14). To grasp what happened at the cross, we must also make trinitarian distinctions in the activity and nature of God.[19]

Further, the cross cannot be separated from the resurrection, and the early church perceived the trinitarian agents as also active in this event. Jesus did not raise himself, but the Father raised the Son through the Spirit (Rom. 1:4, 8:11; 1 Peter 3:18). And then the relationships reversed somewhat, for the risen Son received the Spirit from the Father and poured the Spirit out on all flesh (Acts 2:17, 33; John 14:16, 15:26; Mark 1:8 and parallels). In turn, the Spirit began witnessing to the Son (John 16:13–15), urging all creatures to confess that "Jesus is Lord" (1 Cor. 12:3, Phil. 2:10–11).

Christians have regarded the events just described as the source of human salvation. Yet they can be perceived within a broader context: the apocalyptic sufferings and renewal of all creation. The sun's darkening and the earthquake that accompanied the crucifixion indicated that nature was being incorporated into this travail. Moreover, the Spirit that raised Jesus bodily and began groaning within humans also began groaning and surging toward newness throughout all creation (Rom. 8:11, 18–23). In this kind of "apocalypticism," however, the material world is not destroyed to make way for a spiritual one. Instead, the forces of death and life struggle to pervade human and nonhuman creation.[20]

This understanding of the redemptive work of Jesus emphasizes the cross. Here the entire Godhead entered into suffering and death and at the same time bore creation's apocalyptic sufferings. Redemption consists largely in the divine Reality throwing itself open in this way so that creatures may participate, an action that can be called *divinization*. In this Reality becomes intimately present in their suffering and joy.

While such an understanding of redemption may sound unfamiliar to Mennonites (and many other Christians), it is deeply rooted in the concrete activities of Jesus' life, death, and resurrection. Its trinitarian character flows not from speculations about essences and persons, but from the very shape of this history. It understands the concern of Jesus for the victimized (including nonhumans) and his *shalom* vision as ethical teachings, and as more. For these were also steps through which God, on the way to the cross, opened Godself to identify with creaturely suffering and to invite creatures to share its life and joy. Moreover, this process by which everything comes under the radical, living lordship of Jesus enables one to affirm something about creation. Let us see how.

IMPLICATIONS FOR CREATION

In the Old Testament, creation is understood in light of the redemptive activity of God. The same is true in the New Testament. Jesus' redeeming work is frequently called a new creation;[21] it is presented as the culmination of the plan for which God designed the cosmos.[22] Consequently, theological knowledge about creation has to do with how God is creating the universe anew through Christ. It also involves retrospective consideration, in light of this fulfillment, as to what the original creation and its purpose were like.

Christians also believe that, in the redeeming work of Jesus, God was revealed as God actually is. God's character was not manifested indirectly, as it had been through earlier prophets. Instead, through Jesus, God entered history in person and revealed Godself as God truly is.[23] Accordingly, from this climactic revelation we can infer something about the original and continuing relationship of God and his creation.

THE GOD–WORLD RELATIONSHIP

The issue of the God-world relationship is extensively discussed by more "liberal" environmental theologians. Is it best to consider God's own being as intrinsically intertwined with the being of creatures? Is panentheism necessary or highly desirable to convey that God truly cares about them? Is the traditional teaching unfounded, that the being of God is distinct from the being of creatures? And does this teaching carry opposed environmental implications?

Our focus on the life, death, and resurrection of Jesus shows that God comes very close to creatures and is deeply concerned about them indeed. During his ministry, Jesus shows special sensitivity to the weakest and most inconspicuous humans and nonhumans. On the cross, the sufferings of all creatures are taken up into the Godhead's own experience. With the resurrection, God's Spirit begins awakening a longing within all creatures for fuller participation in Life.

When we approach this indwelling of creatures in God and of God in creatures from the standpoint of redemptive history, however, these appear not as relationships that have existed from the beginning, but as those that result from God's special self-opening and self-giving. That is, the redemptive closeness of God to creatures seems not to be a condition existing "by nature," but one that comes to be "by grace." The impression that God approaches us voluntarily, by grace, heightens when we remember the fierce opposition Jesus faced. God did not come to those who welcomed his advent, but to those who resisted it so violently that they killed his Son. Yet God's love kept flowing out toward them, even through the pain of rejection, abandonment, and death. God died for his own enemies, as Anabaptists have stressed.

If, as the pantheists say, creatures have always formed part of God's "body," then his suffering for and with them would not be wholly gracious and voluntary, for one inevitably experiences pain when a member of one's body dies, simply because it is an inextricable part of oneself. God's love for such a member would be what theologians call *eros:* where one loves something because it contributes to or enhances oneself. But if no natural interconnection between God and creatures exists, then God's love for them is wholly voluntary, gracious, and what theologians call *agape:* where one loves wholly for the benefit of the beloved object.

Agape is the kind of love that energized Christ's redemptive activity (esp. 1 John 4:9–10).[24] It was most radically exercised when Christ died for those who not only had no natural connection with God but were God's enemies (Rom. 5:10).

If we consider the God-world relationship in light of God's gracious redemptive work, then God and creatures seem to have always been distinct. This confirms the impression given by Genesis: that God called creatures into being simply through the divine word and that they did not arise from any preexisting material that might have formed God's "body." Genesis underscores this by placing the little-used word *bara'* at crucial points (Gen. 1:1, 21, 27; 2:3–4; 5:1–2). In contrast to other verbs of making, which imply the use of preexistent materials, *bara'* indicates that something is created directly in its entirety.[25]

The New Testament affirms this same point, but in light of Jesus' resurrection. This raising of Jesus out of apocalyptic death throes and abandonment was regarded, as we have seen, as the beginning of a whole new creation. Therefore, Paul could call God the one "who gives life to the dead and calls into existence the things which do not exist."[26] Elsewhere, he could say that God chose what is low and despised in the world and even "things that are not, to reduce to nothing the things that are" (1 Cor. 1:28). Here is another instance where God's relationship to creation emerges most clearly in light of the process that brings everything under the radical, living lordship of Jesus.

To affirm that God's being is *distinct* from creation's is, as we have seen, by no means to affirm that God is *distant*, as panentheists often suppose. If God takes the sufferings of creatures and indwells them out of grace, God comes astoundingly close to them, and God's love for them is greater than it would be if they were naturally interconnected with God. Panentheism minimizes the width of the gaps spanned by God's love, touching and suffering not only for creatures who are essentially different from God, but also, as Anabaptists stressed, for those who inflict violence, hatred, and enmity upon God.

THE ORIGINAL CREATION

If God was revealed as God truly is through the life, death, and resurrection of Jesus, then the trinitarian distinctions revealed there must somehow characterize God's reality. The historical relationships among these three agents were not hierarchical, but ones of mutual love and cooperation.[27] From this we can infer certain features of God's original relation to creation. (I realize that I may be moving, for lack of space, more swiftly from the historical to the suprahistorical than some Mennonites will find comfortable.[28] Even though trinitarian reflection may be unfamiliar to many Mennonites, the notion of God as Trinity, or inherently communal, provides the strongest possible foundation for the Mennonite emphasis on community.)

The eternal trinitarian relationships, based on what we know of the historical relationships, can best be described as those of mutual love, adoration, cooperation, and self-giving. All this has been called *perichoresis:* a continual, mutual, dynamic interchange of these energies among the persons. Within the eternal Trinity, as in its redemptive activity, the divine *agape*-love and energy is always going out of itself, giving itself for an other. Based initially on what we know of redemption, then, it is appropriate to think of the cosmos originating from an overflow of this perichoretic *agape.* God desired that others should share in the adoration, cooperation, and joy occurring in God's own life. And so God created a universe that would mirror, to some extent, the divine *perichoresis.* We can regard the interconnectedness among creatures stressed by panentheists not as evidence that God is ontologically/actually one with them by nature, but as an image of this trinitarian interrelatedness.

We have also noticed, however, that God's redemptive approach to creatures involved a self-humbling, a self-limitation. And though the images of "overflow" and "limitation" seem to clash, we can also think of God's relation to creation as involving the latter. (In speaking of such matters, we must employ metaphors, which can never be perfectly consistent.) If the cosmos has not always existed alongside God, as something like God's body, then the original creation must have involved a self-limitation. For before creatures came to be, God was the only reality there was. Creatures could emerge, therefore, only if God opened up a space inside herself, as it were, where this could occur. But in so doing, God would limit, and humble, Godself, allowing creatures to exist in a free space within her.[29] Seen from this perspective, God's self-humbling in Christ and the indwelling of the Holy Spirit were not unusual, emergency events undertaken solely for human salvation. They were, rather, further expressions of the attitude that God adopted from the beginning toward the whole creation.

As genuine counterparts of God, as products of *agape* and inhabitants of an open space, it is not surprising that some creatures, at least, are gifted with freedom. Freedom makes possible a full, deliberate response to God—but also, sadly, a turning away. Thus, while God created the conditions in which evil could arise, creatures' free choices are responsible for evil's actual existence. Evil, that is, is distinct from God. In panentheism, however, where all things are part of God's being, evil must also be included. This makes it difficult to energetically oppose evil, environmental or otherwise, because God's own self includes it and seems to sanction it. The Anabaptist/Mennonite tendency clearly to distinguish good and evil, at least on some matters, and resolutely to oppose the latter is better grounded in a God who is distinct from creatures and allows them true freedom.

Further, if creatures are part of God, then God's love for them is conditioned by the fact that they thereby contribute to or enhance God (*eros*), and any freedom they have is significantly limited by this. For whenever anyone loves me largely out of their need for me, there are "strings attached." But if God is distinct from creatures and is a triune community of love, God does not *need* them (though God will deeply *want* fellowship with them), and they are truly free to be themselves and freely choose for or against God.

EVOLUTION

Many "liberal" environmental theologies attempt to derive ethical guidelines from the evolutionary process. Careful examination of evolution's course, they claim, is the main means for learning how we ought to treat our environment today. Over the last few decades, the perspective of many scientists on evolution has altered. Evolutionary theory used to focus on competitive struggle and "survival of the fittest." Today, however, the role of cooperation among species is more fully recognized. This is partly due to the growing influence of ecology, which stresses that mutually supportive interactions among members of ecosystems enable them to flourish and survive. Many environmentalists have been impressed by the Gaia hypothesis, the theory that the earth itself, or at least its living creatures (*biota*), forms a single, interconnected, self-regulating, self-balancing organism.[30]

From this emphasis on interconnectedness, environmental theologians infer, first, that we humans ought to be humbly aware of our dependence for life itself on myriads of other creatures and ought to cease dominating them for our own purposes. Second, since interconnectedness shows that every creature, including those that seem insignificant, is essential for the survival of many others, we should have compassion on all of them. Fundamentally, "the entire insight on which compassion is based is that the other is *not* other; and that I am *not* I . . . In loving others I am loving myself and indeed involved in my own best and biggest and fullest self-interest."[31] Third, most environmental theologians argue that evolution has advanced toward the production of self-conscious organisms and that humans represent a turning point in the process: Through us, evolution has become conscious of itself, and we are now responsible for taking over its direction.[32]

Mennonites ought to welcome the first theme, for it extends the attitude of humility and servanthood, which Jesus taught and lived, to the nonhuman realm. What environmental theologians gain with this first emphasis, however, seems taken away by the third, which exalts humans extraordinarily, entrusting them with direction of the cosmos. In contrast, Anabaptist/Mennonites have believed that human responsibility consists chiefly in following what Jesus taught, leaving overall guidance of history and the universe up to God.

In addition, great difficulties attend attempts to show that evolution manifests any clear patterns of "advance," such as that of consciousness.[33] Even if evolution has advanced, it has done so not chiefly through maintaining ecosystemic balance, but through episodes in which some or many species have been annihilated to make room for new ones. Cooperation, indeed, may have played a greater role than heretofore recognized in enabling species to flourish. Yet current evolutionary theories still assume the Darwinian premise that new species appear, for the most part, when older ones decrease or become extinct. Enough destruction is attributed to evolution, even in recent theories, to make it a poor guide for environmental ethics.

Although contemporary ecosystems, in contrast to the disruptive evolutionary process, may seem to provide a strong basis for cooperation, even they cannot support an ethic of compassion for the weakest.

In an ecosystem, it is precisely the weakest, in the sense of the least robust, that are repeatedly consumed for the good of the whole. Nonhuman nature, whether in past evolution or current ecosystems, shows no real concern for individuals, especially if they are weak or ill. Even the argument for compassion based on interconnectedness is weak. First of all, my own self-interest is not really dependent on *all* other creatures; I could abuse many of them and still live quite well. Second, the argument's appeal is precisely to *self*-interest, to an *eros* kind of love where I value others because they will benefit myself.

Our current environmental crises, however, require an ethic that goes well beyond self-interest, one that is able to value other creatures in and of themselves, even when they have no apparent connection with one's own well-being. Only the conviction that all creatures have some intrinsic value can adequately check the tendency to use them solely for our own ends.[34] Such a concern for even the weakest and apparently most useless creatures, however, cannot be derived from evolution or even from ecosystemic balance. It is clearly rooted in the concern and suffering of Jesus for even the least of creatures, which shows that God values them much differently than these current sciences suggest.[35] This perspective also underlies Jesus' emphasis on nonviolence.

ESCHATOLOGICAL DESTRUCTION?

Whereas some "liberal" theologians conceive God as so closely related to creation that creation is God's "body," some "conservative" Christians regard God as so unconnected with creation that He will one day destroy it. To be sure, some New Testament texts seem to imply the latter. But let us now see whether the theology of creation that we are outlining can help us gain a broader, more cosmic perspective on this issue.

Creation, I have posited, originated in an overflow of divine love. Its many interconnections mirror the glory of the divine *perichoresis*. It hardly seems possible that God would have little regard for a cosmos created in this way. Of course, creation's physical dimension might possibly be a "lower" reflection of God, to be superseded some day by a "higher," spiritual one.[36] Yet we noted that God took physical creatures seriously enough to be limited by their reality and gave at least some a freedom by which God could be affected negatively.

Eventually, God the Son took on a body such as all humans have. Thus, this Person of the Godhead became as intertwined with all other creatures as other humans are. This incarnation, however, was no wholly isolated emergency measure. It was consistent with the attitude toward creatures adopted in God's original self-limitation. During his ministry, Jesus Christ showed great respect for apparently insignificant nonhuman creatures, indicating that such creatures have intrinsic value in God's eyes.

At the cross, the sufferings not only of humans but the apocalyptic sufferings of creation as well were borne by, and directly affected, the Father, Son, and Spirit. In allowing itself to be afflicted by creatures' finite, physical pain, the Godhead regarded their material form with extreme seriousness. The Son then arose from death in a form which, while it transcended ordinary human experience, was in some basic sense a body. Yet this was but the "firstfruits" of the coming resurrection of his people (1 Cor. 15:23). Since Scripture describes this event also as bodily, it seems to imply that in it humans will retain some interconnection with nonhuman nature, presumably through some resurrection of the latter. The risen Jesus, through his transformed body, must also retain some significant connection with the nonhuman realm in the present, though it is difficult to specify how.

Through the Holy Spirit, however, God's connection with nature is not only maintained but intensified. The Spirit makes the human body God's temple (1 Cor. 6:19) and renews it inwardly with a groaning that parallels the Spirit's groaning throughout all creation (Rom. 8:11, 18–23). Indeed, in this kind of travail toward renewal and in the corresponding sufferings on the cross, a genuine "apocalyptic" element appears—yet one that looks forward to Earth's transformation, not its destruction. Since the Spirit arouses hope and longing for the fullness of God's new creation, yet works intimately within bodies in so doing, it hardly seems possible that the Spirit could be urging those bodies toward total destruction.

It is more appropriate to understand texts that might seem to predict coming destruction (Luke 21:9–28; 2 Peter 3:10–13) as pointing dramatically toward a final transformation that creatures will undergo in order to fully bear God's new Life. In general, the redemptive pattern throughout Scripture speaks not so much of humans going to heaven as of heaven coming ever more fully to earth.[37] It is appropriate to image, with the last chapters of Revelation, the New Heaven and New Earth not as wholly disconnected from the present earth, but as coming down upon the latter and transforming it through God's presence.[38] This means that whatever we do now to preserve and enhance our environment will in some way be preserved and transformed in the final state.

CONCLUSION

The Anabaptist/Mennonite emphasis on living fully in God's kingdom while on Earth implies a positive valuation of creation, for that life is carried on in the material world. Yet the early Anabaptists, due to their intense preoccupation with ecclesiastical and social evil, never worked out this implied theology of creation. Contemporary Mennonites, therefore, may appropriately seek to articulate this missing piece.

I have begun from what I regard as the focal point of Mennonite theologizing—the process that brings everything under the radical, living lordship of Jesus Christ. If Mennonites begin simply with Jesus' life and teachings or with a particular teaching like peace, they will not be likely to attain the cosmic perspective required for addressing environmental concerns. The process that brings everything under Jesus' lordship includes his life, death, and resurrection and creation's subsequent renewal up to the eschaton (the end of history as we know it). It begins, however, with a theme Mennonites have always stressed: the way of Jesus to the cross.

Examination of this process shows that it must be regarded as the cooperative work of God as Father, Son, and Spirit. From the main features of this redemptive process, theology can infer something about the relation of God to creation. From redemption's utterly gracious character, which involves God dying for God's enemies, theology can infer that creatures have always been distinct from God and not intertwined with God's being, as panentheistic environmental theologies claim. Panentheism also tends to undercut true creaturely freedom and to include evil in God. Theology must also derive its guidelines for environmental involvement from the self-giving activity of God in redemption—which values the least of creatures and is consistent with nonviolence—and not from evolution, as many environmental theologies propose. Finally, God's close, revitalizing relationship with creatures throughout creation and redemption indicates that our present Earth will not be destroyed, but will be transformed, at the eschaton.

CHAPTER 11

The Earth Is a Song Made Visible

Lawrence Hart

Native American perspectives regarding the earth may offer cues and influence actions as one considers alternative lifestyles to counter pollution and destruction of the environment. The Native Americans' relationship to the earth is legendary; they have always maintained an intimate, spiritual, and personal relationship to the earth. Mountains, forests, streams, rivers, oceans, and the sky are a part of a circle. All life forms in the waters, on the land, and in the air are interconnected within the circle. Humankind is a part of that circle. This view holds that there is a sacredness about the earth and, when any part of that which is upon the earth is mistreated, all forms of life within the circle, including humankind, are mistreated.

It is less well known or simply unacknowledged that cultural customs regarding the earth practiced by Native Americans have parallels in both the Old and the New Testament. One that immediately comes to mind each time I see a Cheyenne traditionalist touch the earth ceremonially is Psalm 24:1. The psalmist declares that

The world and all that is in it belong to the Lord,
the earth and all who live on it are his. (TEV)

The ritual of a Cheyenne traditional priest touching the earth four times is an acknowledgment that the earth is a creation of God and a part of the circle of life. In a matriarchal family system, a system predominant in Native American cultures, it is readily acknowledged that the earth is the mother of all living things.

NATIVE AMERICAN VIEWS REGARDING THE EARTH

Beliefs or views Native Americans hold with respect to the earth either directly conflict or have varying degrees of differences with views held by the majority of other peoples in North America. These views are pertinent to any discussion of Native American perspectives on the environment. In the preface to *The Encyclopedia of Native American Religions*, Arlene Hirschfelder and Paulette Molin state that "Native American sacred beliefs are as dignified, profound, viable and richly faceted as other religions practiced throughout the world. Native sacred knowledge has not been destroyed or lost but in fact lives on as the heart of Native American cultural existence today."[1]

A brief focus on four such beliefs will suffice to highlight differences and point to similarities. First, there is still a strong people-hood concept among Native Americans throughout this hemisphere. All thought and action begins from the first person plural. As an individual I may use the personal pronoun. However, I can never say *I* without being cognizant that I am a Cheyenne. My identity as an individual Native American is connected to and deeply rooted in a particular tribe. Individualism is a foreign concept to Native Americans. An individual Native American born into a tribe will immediately have an identity with a subgroup, which may be an extended family, a clan, or a society of the tribe, that continues throughout life. The tribal subgroups have a function to benefit the whole. If a group fails to perform a function, the whole suffers. If it succeeds, the whole benefits. The people-hood concept held by Native Americans conflicts with the rugged individualism especially espoused in the United States.

Second, and closely tied to people-hood, is the Native American concept of relationship to the earth. A Native American does not hold an individual view about the earth but instead a group or tribal view. The earth is for the whole circle of people. Individual ownership of any part of the earth is a foreign concept. The earth belongs to the Creator, and there are numerous accounts in literature of Native Americans objecting to the practice that the earth is property to be owned and sold. According to Sharon O'Brien, an associate professor in the Department of Government and International Studies at the University of Notre Dame:

> Despite their debts to Indian culture, European's treatment of Indians was generally hostile and always self-serving. The pattern varied from virtual extermination by the Spanish to hostile dismissal by the English to grudging respect by the French. European civilization was based on individualism, hierarchy, and materialism, and Europeans considered their way vastly superior to Indian cultures. Reared in societies that emphasized acquisition through competition and control, Europeans were simply unable to appreciate or even understand cultures that de-emphasized those values.[2]

Land ownership by individuals was deemphasized, and this position is still held by the older generation. However, each group or tribe has always felt closely tied to that part of the earth on which they have been placed to dwell by their Creator.

A third and central view held by Native Americans is that the earth has a sacredness about it. A parallel view in the Old Testament is that the earth is a creation of God. As with the Cheyenne priest first touching the earth, one cannot conduct an action or speak without first acknowledging the earth to be a part of being. A ceremony or a ritual must precede an activity that is to be conducted by any subgroup or the entire tribe. And the ritual is performed by a priest. Such a ritual is tantamount to an invocation. To conduct an activity or to speak without the invocation is a desecration of that which is sacred.

NATIVE AMERICAN COMPARED WITH ANABAPTIST VIEWS

These foregoing world-views—namely, people-hood, closeness to the land, and the sacredness of the land—can be appreciated by most religious groups. Anabaptist and Pietist groups especially would have an affinity to the concept of people-hood, although its acceptance and practice might vary in degree. The Anabaptist tradition of closeness to the land is long lasting and is unequaled in other ecclesiastical groups. For those religious traditions practiced in rural areas, closeness and affinity to the land will not be strange.

However, a view that the earth is sacred is not accepted. As an Anabaptist I appreciate this, for I am aware that the early Anabaptists took a position different from that of other Reformers regarding the earth, as this book testifies. Through a letter referred to as "the charter of the free church," they stated that sacredness does not attach to special words, objects, places, persons, or days.[3]

It has been well known that the aboriginal people of this hemisphere had a strong connection to the land. It is less known that they were outstanding agronomists and horticulturists. Native Americans tilled the ground, fertilized it with products from the earth, and planted seeds that they selected with care. The ensuing harvest enriched their life and culture. Just before the Columbian quincentennial in 1992, much was written on the many contributions the aboriginal people made to the world. For centuries Native Americans tilled the earth and planted seeds. They

> produced over three hundred food crops with many having dozens of variations. The people from the Old World gradually transplanted many of these crops from America, and each in turn contributed in various ways to improving the world diet in both quantity and quality of foods. The Indians gave the world three fifths of the crops now in cultivation. Many of these grew in environments that had formerly been inaccessible to agriculture because of temperature, moisture, type of soil and altitude.[4]

RITUALISTIC SINGING AND RESPECT FOR NATURE

Another Native American practice relating to the earth has to do with singing. Conspicuously absent from written accounts of Native American culture are references to ritualistic singing. All activities in Native American life involve a song or songs, and this is especially true of the Cheyenne people. In the introduction to *Southern Cheyenne Women's Songs*, Virginia Giglio explains that "the Cheyennes place a high value on music . . . Music is a gift considered to be from a spiritual source."[5] Songs in all likelihood have lyrics about the earth. This tradition is still prevalent within Native American church life today and is still practiced in cultures where Native Americans have not been swayed by strong forces of acculturation and assimilation, including influences from some quarters of the church.

In preparation for tilling, a common practice by many Native American tribal groups was to sing a song before the earth was broken, a custom valued because tilling the earth is a God-ordained task. Too often we think of tilling as laborious and mistakenly accept the notion that it is a consequence of disobedience. We must remember that tilling the earth was God-given before the fall (Gen. 2:15).

Yet another song was sung before planting seeds, and there was a song for the harvest. These were common rituals. Indeed, for every occasion in Native American life, there was a song—songs at birth, songs for life's activities, and songs at death. Singing of songs by Native Americans in significant times is no different from the practice of the Hebrew people in both the Old and the New Testaments. When I hear songs sung today, especially by Indian women, I am reminded of the song sung by Miriam after crossing the Red Sea, that of Hannah when God granted her wish for a son, the Magnificat sung by the chosen Mary, or songs sung when the early church gathered.

When Christian missionaries first arrived, the Pueblo people still sang songs before planting the special varieties of corn they had developed. New converts were expected to cease practicing this and other cultural traditions and thus, invariably, there were cultural conflicts. Cultural conflicts still exist on that reservation today. Some Christian missionary workers have suggested that the culture of these people should be viewed as analogous to that of the Old Testament period. Christian workers and Pueblo Christians could presumably build upon the Old Testament culture and add to it the Good News of the New Testament. Unfortunately, this excellent recommendation has yet to be followed.

Ritualistic singing is still practiced by many tribes and reflects the strong connectedness to the earth that was shared by all tribes. The singing of songs, especially about the earth, is current practice in and outside the membership of Christian churches. Many songs in the Cheyenne language expressing their understanding of the Christian faith reveal tribal traditions of connectedness to the earth. In many songs from the birth of Jesus to his return—whether they are translated hymns or indigenous songs—there are references to the earth. In fact, the hymns gain greater meaning when the message in the song has lyrics about the earth.

An English version of a hymn may not have a single reference to the earth but, if the translator desires effective communication with Native

Americans, new lyrics should be composed and the word *earth* added. The end result will be not a literal translation of the hymn, but one culturally specific and relevant. Moreover, the message will gain profound meaning.

A classic example is *Silent Night*. This well-known and internationally used hymn composed in 1818 has no reference to the earth in the original German nor in the English translation. Rodolphe Petter, a Swiss linguist who studied Cheyenne culture and language in Oklahoma and Montana at the turn of the century, captures the rich meaning of the incarnation. He employs the word *ho 'e va* (earth) in translating this hymn.[6] The Cheyenne translation is a marvel. It accurately states the theology of the incarnation by referring to the earth. Perhaps Petter's Cheyenne language informant should take compliment. Because of the use of *ho 'e va*, this hymn, which communicates the birth of the Christ child, has a much deeper and more profound meaning for a people whose culture is inextricably linked to the earth.

Many indigenous Christian hymns and songs Native American Christians have been inspired to compose abound with the word *earth*. An indigenous song composed by a Cheyenne peace chief on the subject of eschatology, with the translated English title of "Someday," has the word *ho 'e'* (earth).[7] The message of a coming rapture of the earth is effectively conveyed.

A Cheyenne Indian educated at the Carlisle Indian School in Carlisle, Pennsylvania, composed a Cheyenne song that is now sung for praise and adoring by Anabaptists and Pietists scattered throughout the world. A Christian convert who was one of the language informants to Rodolphe Petter, Harvey Whiteshield obviously appreciated his culture, for the song he composed has a reference to the earth.[8]

Because he is a role model for contemporary peace chiefs, I continue to be impressed with the final song sung by Cheyenne peace chief White Antelope. As the Sand Creek Massacre unfolded on November 29, 1864, he stood in the middle of the village, unarmed, and sang his death song.[9] According to our oral tradition, the words are

> *Father have pity on me, Father have pity on me,*
> *The old men say, only the earth endures.*
> *You have spoken truly. You have spoken well.*

Nothing lives long, only the Earth and the Mountains.
Nothing lives long, only the Earth and the Mountains.

Is there a parallel to White Antelope's song in the Old Testament? Yes, the first part of Psalm 90. The Psalmist refers to the finiteness of succeeding generations of humankind. In further juxtaposition the Psalmist refers to the mountains, the earth, and the world (RSV).

Another story of a song is worthy of mention. When the troops of the Seventh Cavalry under the leadership of Colonel George Armstrong Custer attacked a peaceful Cheyenne village on November 27, 1868, no one had time to sing a song. Oral tradition relates the story of a woman who had escaped the initial attack by running from the village along the Washita River. Upon reaching a knoll, she turned to see her people, including children, women, and elders, being killed by the troops, who had dismounted and taken a position after the initial charge through the village. In the surprise attack at dawn on a cold, snowy morning, there had been no song sung, for all was chaos. This woman began to sing a song for her people. Custer's troops tried to silence her, but she stood on that knoll and continued singing, for she was beyond the rifle range of Custer's sharpshooters. In all the inquiries I have made, no one can recall the words sung by this courageous woman. They may be lost to memory, yet it is possible that an elder knows the words and will reveal them at an appropriate time.

When this oral tradition account was related to the Little Big Horn Associates, whose members are admirers of Custer and the Seventh Cavalry, a woman not associated with the group but in the audience was moved to write a poem. It was completed by the time I finished the story, and Anne Clement, an ordained Methodist minister, gave me her composition, entitled "And the Woman Sings."

In the time of the
Red Moon her song is sung.
From the heart it comes
without words understood
only with the spirit.
In the midst of death
life sounds across the red hills

carried by the winds.
Somewhere the death song
becomes the sound of life
and the song goes on and on and on.
All the sounds of death cannot silence
one note sung for life.

These words could be lyrics for a new song.

Speaking to members of the school board, school administrators, and educators at the 1996 annual convention of the National School Boards Association, the national teacher of the year, who teaches in a small Native Alaskan village on Kodiak Island, related a story about a basket she had woven under the tutelage of a female Alaskan elder. After completing her first product, Ann Griffin commented that her effort resulted in a basket not especially noteworthy. Her elder instructor, a first teacher of Inuit tradition, disagreed and taught this educator "a basket is a song made visible."

NATIVE AMERICAN CONTRIBUTIONS TO THE PROTECTION OF THE EARTH

These two stories reveal positive influences of Native Americans on people of other cultures in recent times. They are shared to illustrate the prospect that Native American views of the earth can provide similar influences to counter pollution and destruction of the environment. Perhaps we can have a second discovery of America by listening to the voices of Native Americans.

A distinguished Canadian former jurist, Thomas R. Berger, writes in *A Long and Terrible Shadow*,

> The culture of Native people amounts to more than crafts and carvings. Their tradition of decision-making by consensus, their respect for the wisdom of their elders, their concept of the extended family, their belief in a special relationship with the land, their regard for the environment, their willingness to share—all these values persist in one form or another within their own culture, even though they have been under unremitting pressure to abandon them . . . This is still the age of discovery, a discovery of the true

meaning of the history of the New World and of the Native People's rightful place in that world. This is a discovery to be made in our own time should we choose it—the second discovery of America.[10]

Just a few years ago, a Christian anthropologist was commissioned to study the cultural conflicts between people in the church, both non-Indian Christian workers and converts, and a traditional Pueblo people on their reservation. Historically, it had been expected that the converts would put off all their cultural practices, for they were seen as pagan. Ironically, these were the very people who, through their agronomy, had produced a variety of corn whose stalk was short in height but produced a long cob that could reach maturity in a few weeks. The Pueblo people produced many varieties, including the famous blue corn. The varieties of corn developed could withstand arid conditions and, with little moisture in a short time, produce more than enough corn to meet their needs.

According to the account in the first chapter of Genesis, God created the heavens and the earth. The earth was without form and void, and darkness was upon the face of the deep. God then created all that is of the environment and pronounced it good. Then He created humankind and pronounced this final creation as very good. Using a Native American perspective, one can view the created earth as a song made visible. The earth God created and declared to be good should inspire us to sing.

The song we sing about the earth can have beautiful lyrics. It may express gratitude for what God has accomplished, just as when the Hebrew people crossed the parted Red Sea on dry ground. After crossing and gaining freedom from oppression, they sang a song of thanksgiving. It has beautiful words. Miriam was inspired to add her song, and it, too, is beautiful: "Sing to the Lord, for he has triumphed gloriously" (Exod. 15:21a RSV).

I have been encouraged that our worship liturgies have changed over the last three decades. Language and lyrics are now inclusive. Women in the church have led this change. More recently, ritual is gaining acceptance and usage. More references to the earth should be made in music compositions, and I trust our talented composers will receive inspiration and be empowered to perform the task.

In our treatment of the environment, we have a choice between the sacred and the profane. If we view the earth as a sacred gift of God and treat it accordingly, we, too, will sing a song with beautiful lyrics. On the other hand, we can mistreat the earth by continuing to participate in pollution and destruction. If we choose to continue to mistreat the earth, the lyrics we sing will be as profane as most of today's rap music. Profane music is auditory pollution, and it is a sign of the times. Pollution and destruction of the earth are profane.

This book can help us gain or deepen our respect for God's good creation and allow his Spirit to empower us to a commitment to treat the earth in alternative styles and methods. Such styles and methods may lead us to actions that seem radical. We can no longer wait. Ours may be a calling to stop the pollution and destruction of the good earth with radical measures. If we so choose we can be radical. After all, we are Anabaptists.

"The earth is the Lord's."

And the earth is a beautiful song God made visible.

PART IV

The Challenge to Take Care
of the Earth

CHAPTER 12

Toward an Anabaptist/Mennonite Environmental Ethic

Heather Ann Ackley Bean

The Anabaptist record regarding environmental issues is rather ambiguous, as the previous chapters have shown. In addition, until recently, little had been written upon which to develop an Anabaptist environmental ethic. Therefore, testing the record of Anabaptist ecological writing and practice against Anabaptist theology and ethics reveals both problems and promise. But theologian James McClendon gives us direction by proposing that the task of an Anabaptist environmental ethic "is simply to show that our creaturely existence is morally legitimate" and a cause for "natural delight."[1]

INCONSISTENCIES AND PROBLEMS IN ANABAPTIST THEOLOGY

The global context to be considered when developing this ethic is *ecocide*—the slow but certain death of entire biosystems due to human wantonness, especially in North America.[2] An analysis of Anabaptist theology involving environmental ethics must begin by recognizing that, historically, environmental issues as we understand them today were not an Anabaptist priority (which is also true for most other Christian traditions). The relationship between Anabaptism and its environmental ethics is complex. After a search of "the Anabaptist tradition for traces of environmental concern," Mennonite theologian Walter Klaassen concludes, "Despite a commitment to nonviolence, Mennonites . . . have done no thinking about nonviolence toward the Earth and 'are by no means in the Christian front ranks of creation care.'"[3] Agreeing with Klaassen's conclusion, Redekop confirms that "there is absolutely no reference to the preservation of the earth in Mennonite theology."[4]

Anabaptist theology itself is partly responsible for this oversight. The Anabaptist response to the increasingly serious ecological crisis comes out of this dualistic theological and ethical perspective. As Finger indicates, the world belongs either to the kingdom of God or the "world of Satan."[5] In Anabaptist dualistic theology, the church of believers under the rule of Christ must separate from the world, the kingdom of Satan. "The problem was that 'the world' and the natural nonhuman 'world' were not conceptually or existentially separated, and thus in the process the nonhuman creation became identified with the evil in the 'world' from which the pure were to abstain."[6] Further, the human "world" outside the church as a community of believers was never distinguished from the rest of creation.

Early Anabaptist groups shared the "kingdom ethic" born of their two-kingdoms theology and thus differed from their contemporaries. Inspired by the emphasis on reconciliation and social transformation in the Sermon on the Mount and Matthew 18, the original Anabaptists attempted, both individually and socially, to practice the Gospels' teachings in daily life. Their interpretation of these Scriptures created an Anabaptist belief system centered on "concern for community, radical discipleship, literal adherence to the Sermon on the Mount, non-resistance, and nonconformity."[7] A truly Anabaptist environmental ethic should be based on these early common values.[8]

Theologian Tom Finger subsumes these ethical principles to the primacy of Anabaptist belief in the absolute redemptive reign of Christ.[9] While maintaining the importance of Christian redemption in Anabaptist thought, the focus on dualistic biblical teachings tends toward antiworld, anthropocentric thinking. The absence of an explicit creation doctrine compounds this problem. The standard Anabaptist interpretation of the New Testament characterizes the world and flesh in direct opposition to the realm of the divine. To many Anabaptists, the Bible is clear: "Love not the world, neither the things that are in the world" (1

John 2:15).[10] If this verse is taken to summarize the Bible's teaching on Christian responsibility for the created order, the absence of an Anabaptist environmental ethic is understandable. Appealing to the apostolic authority of Paul and John, the writings of Menno Simons himself exhort Anabaptists to "set your affections on things above, not on things on the earth . . . Dwell in the heavenly reality and appear to the world no longer to live."[11]

This biblically based dualism caused early Anabaptists to anticipate God's triumphant destruction of the world "along with the evil, unregenerated human order," according to Finger and Redekop. Such a view is inhibiting if not antithetical to active concern to prevent ecological crisis. Modern Anabaptist theology focuses on "biblical nonresistance" rather than two-kingdoms theology, but, like other Christian theologies, its concerns are often so abstract as to seem unrelated to practical contemporary issues. In addition, Anabaptists' "unexamined or static notion of nonresistance has sometimes led to an attitude of simply expecting the government to do the fighting and protect the people."[12] This common approach encouraged the avoidance of personal and denominational responsibility for environmental action, since that belonged to the world and the state.

The lack of a creation doctrine is not a mere oversight in Anabaptist theology but arises inevitably out of its anthropology. Anabaptist biblical theology leans more toward the creation account in Genesis 1, wherein humanity is center stage and dominates nature, than the seemingly contradictory account in Genesis 2, which suggests that humanity is the servant of God's creation. The Pauline interpretation of these accounts, which further downplays Genesis 2's emphasis on creation to focus solely on the human relationship to God, was authoritative for Menno Simons and has deeply influenced Anabaptist thought in general. Even James McClendon's attempt to lay the foundation for an Anabaptist environmental ethic in his *Systematic Theology: Ethics* is unapologetically anthropocentric: "Christian ethics [should] focus first on the *embodied selfhood* of the human species, seeking to understand our bodily nature both as the consequence of the natural history of *Homo sapiens* (that is our link to the environment to its far limits) *and* as the locus of an interiority of shame, delight, guilt, and virtue rooted in the narrative tradition that is ours."[13]

Thus, in the most thorough contemporary Anabaptist systematic theology and ethics, human selfhood is proposed as primary for Christian ethics of the body and of creation. McClendon acknowledges that constructing a creation-centered (rather than human-centered) ecological ethics is especially difficult, since "the issues ecological morality confronts are most often ones that involve not natural but social conflicts" with "governments, multinational corporations, the attitudes of racial groups and interests and nations" and therefore "cannot be resolved in terms of nature alone."[14] His analysis suggests that an Anabaptist ecological ethic may have to be anthropocentric, in the sense of being centered on an understanding of human responsibility for just relations with the created world.

McClendon reflects this self-critically anthropocentric Anabaptist ecological ethic when he decries the "widespread but shameful depreciation of our fellow creatures the beasts." Although McClendon suggests that Mennonite ethics presupposes a doctrine of creation, Redekop argues that this doctrine has never been sufficiently articulated, partly because "the awareness of the limited nature of creation and of our dependence upon it was not as pertinent" during the Reformation, but mostly because Anabaptists' attention was focused on their persecution and martyrdom.[15]

Anabaptist theology has been impeded by this understanding of the created world as the realm of Satan separate from God and believers, by anthropocentrism, and by the absence of a clear doctrine of creation. These impediments have led to some problematic environmental practices among Anabaptists. Although such theological dualism has historically inspired Christian theology's distrust of the created world, Anabaptist kingdom ethics can provide a positive basis and resource for constructing a green theology and ethics.

INCONSISTENCIES AND PROBLEMS
IN ANABAPTIST PRACTICE

A critical evaluation of the development of an Anabaptist creation ethic requires a survey of inconsistencies and problems in Anabaptist *behavior* toward the environment. Historically, Anabaptists have neglected environmental issues, not only in their theology but in everyday practice. For example, despite strong historical ties to agriculture and the land, "Mennonites have offered environmental causes relatively modest support."[16] The reason may be that stated by Art Meyer: "The primary reason that our earth is so environmentally degraded today is that many of us have not really believed that the natural world is that important."[17]

In North America, Anabaptists seem to become less interested in environmental issues and land ethics as their ties to their traditions decrease and they depart from the simple life of agricultural communities to assimilate into the materialistic mainstream culture (see chaps. 5 and 6). Traditional Anabaptist animal-based economies and diets and some Anabaptist service and mission efforts may even have contributed to environmental neglect and damage.

The rising costs of farm equipment and land, the scarcity of land for sale, and spreading urbanization have made it imperative for Amish and other Anabaptist-related groups to farm in more intensive and ultimately environmentally unfriendly ways. A contributing factor is the dramatic increase in the Amish population due to the tradition of having very large families (an environmentally unsustainable practice in itself). According to Levi Miller, who grew up Amish, young Amish men who cannot find farms turn to farm-related business, cottage industries, woodworking, construction, and even industry. Mennonite farming has begun to resemble modern agribusiness, with more environmental threats.[18] The relationship to nature described by Kline in chapter 4 is unfortunately not totally representative of Plain Mennonites and Mennonites in general.

Men's work away from the home and the farm disrupts the agricultural community and unbalances relationships. The move away from agricultural tradition distorts not only human relations but also working relationships with the earth itself. As Marlene Kropf observes, "Modern folk whose days are spent far from lakes and streams and grassy fields may find it more difficult to keep alive such a vivid sense of God's presence and blessing" as that expressed in Genesis and Psalms. Jim Rich, an environmental homesteader, warns that "we have been led to believe that delegating to industry the provision of the material necessities is progress. The result is the gradual weakening of our bond with earth. With our huge power plants and an international food economy our alienation from the earth is nearly complete." The generalizations by Kropf and Rich indict North American Christian behavior in general, but Amish farmer and writer David Kline ties this observation to Anabaptist environmental problems: "Modern Mennonites' inability to find an environmental theology may stem from their alienation from the land," where they now merely reside, instead of "living with it."[19]

Leaving the tradition of farming the land for a living means greater individual economic independence from the agriculturally based church community, which through its self-sufficiency has historically been free to live in nonconformity to the mainstream. Loss of agricultural self-sufficiency has led to an almost inevitable compromise of the Anabaptist values of community, nonconformity to the world, and simplicity. Amish farmer Mose A. Kaufman explains, "It's hard to maintain an Amish lifestyle away from the farm ... It leads to a system of spending ... money we don't have" because "we're in a materialistic society."[20]

The increasing materialism of Anabaptists who have assimilated to mainstream North American values is one of the most significant obstacles to developing an Anabaptist environmental ethic. "Most Mennonites have not followed the Hutterite pattern of establishing and maintaining communal life," a truly ecologically sustainable lifestyle. Although 1990's twentieth-anniversary celebrations of Earth Day increased awareness of the environmental crisis, growing prosperity and upward mobility make it harder for Anabaptists to focus on self-sacrifice, a central value in our martyr tradition. "Awareness and behavioral changes do not go hand-in-hand," Ken Gex observes. "As wealth increases so does consumption and waste of world resources ... Our lifestyles adversely affect others in the world, particularly the poor and powerless."[21]

The resulting prosperity and consumption is in direct conflict with Anabaptist values of simplicity and nonviolence. Prosperity in agriculture can become an obstacle to Anabaptist environmental ethics. One rural minister's wife reports that she and her husband must use the word *environmental* sparingly in their congregation because church members work in "agribusiness" and resist "talk of conservation or sustainable practice." John Oyer suggests that "the Mennonite Church is in danger of becoming merely a mainline denomination." This danger, warns John Ruth, grows as Anabaptists focus on recruiting new converts, increasingly becoming "so 'average' in what we require by way of discipleship [of the new members] that we will end up with little that is distinctive or radical."[22]

Assimilation to mainstream culture not only brings loss of what is distinctive in Anabaptist theology and ethics, but also weaves Anabaptists into the web of corporate sin and responsibility for the environmental damage caused by materialistic living. Assimilation to the materialistic mainstream lifestyle is divisive, threatening Anabaptist communities internally and separating them from neighbors with whom their martyr history should help them identify. Thus they become part of the problem, part of the elite who profit from social divisions such as racism and poverty, intimately related to ecological exploitation (as described in chaps. 1 and 3).

Environmental issues, similar to racial and class differences, further divide Anabaptist groups internally. An African American Mennonite woman cited by Brethren in Christ writer Harriet Sider Bicksler rejects the environmental focus of the Mennonite Central Committee (MCC), charging that "it's too easy for white middle-class people to do a bit of recycling and feel they've done their share" for the environment, "without ever having to 'dirty their hands' by developing relationships with the poor." Bicksler agrees that "whites [including Mennonites] need to broaden their concept of what constitutes a complete environmental agenda . . . Stewardship of the earth ought to include the concern that all people be able to live free of the environmental hazards threatening full and abundant life."[23]

Despite good intentions, however, even the task of how to address such problems divides Anabaptists among themselves and from other Christians, thus preventing effective ecological action. No American Mennonite group is a member of the National or World Councils of Churches, both of which have strong commitments to environmental theologies and ethics. The assimilation of some groups to American consumerism faster than others creates disagreements on how to live in relation to the consumeristic, environmentally destructive mainstream culture. They waste their energies criticizing each other rather than turning them toward environmental healing and creation theology. According to Levi Miller, "The Amish often consider the Mennonites as too worldly, too American . . . Many Mennonites have lost their cultural distinctiveness. The Mennonites sometimes see their Amish brothers and sisters as too separated from mainstream American life, irrelevant, and not evangel-istic."[24]

Further, Anabaptist service efforts often neglect or contribute to en-vironmental damage. White middle-class MCC service workers are, by their own repentant accounts, sometimes paternalistic toward indige-nous peoples in the mission field. MCC volunteer Scott Coats admits, "Outsiders—including me—are constantly talking about environmen-tal protection and forest reserves as though the Akha [Thailand's small-est tribe] have never given it a thought. It's simply not true. They think about it every day. Their livelihood [as forest hunters] depends on it."[25]

Some service projects may even alter the primordial conditions of an area enough that they contribute to the violence done to people by nat-ural forces such as typhoons and floods. For example, using American methods of agriculture, irrigation, and housing development in cultures and climates where they are inappropriate can lead to soil depletion and crop failure.[26] Mennonites are just beginning to become aware of the effects these kinds of environmental issues have on those with whom they work.

Schrock-Shenk reports that, on April 30, 1991, a typhoon struck southern Bangladesh, destroying homes and roads and killing more than 135,000 people. "The dead were among the poorest of all Bangladeshi farmers," who "had moved onto land they knew was especially prone to typhoon devastation, but . . . were too poor to buy land elsewhere . . . Poverty drove hundreds of thousands of people to live in an unsafe place, creating the conditions that turned this storm into a major killer."[27] Al-though Anabaptist service efforts may not have been responsible for the resettlements, service workers are only beginning to take such condi-tions into account as they organize and implement service work abroad.

Mennonite life, traditionally based on animal husbandry, is also problematic for the environment. Gary Comstock, author of *Is There a Moral Obligation to Save the Family Farm?* has been a very outspoken veg-etarian, though also a proponent of traditional family farms. Comstock reckons that Mennonite farms "daily slaughter . . . hundreds of thou-sands of cows and hogs."[28] Vegetarianism, the most ecologically sus-tainable lifestyle, would undermine the economy of many Anabaptist-related families and communities. Thus, even traditional diets and farm practices may interfere with the development of an environmental ethic.

A CONSTRUCTIVE ANABAPTIST RESPONSE
REGARDING THE CREATED WORLD

Anabaptist dualistic and anthropocentric theology and practice are not exclusively detrimental to the environment, however. Anabaptism has borne ecologically good fruit in theory and practice; Anabaptist thinkers have begun to lay the groundwork for an Anabaptist biblical environmental theology and ethic.[29] Milo Kauffman underscores the intrinsic goodness of God's creation and reinterprets his earlier Anabaptist creation theology to develop a more environmentally sound ethic. In his revision, the creator God is omnipotent and omnipresent, while humans, created in God's image, have dominion over and stewardship of God's good creation, which itself is inseparably interrelated with both God and humans.[30]

The late Art Meyer, along with his wife Jocele Meyer, were active pioneer environmentalists because of their Christian commitment and believed that the Scriptures "instruct us to care for God's good earth." They considered the natural world to be "God's creative work" and believed that Genesis 1:31 clearly teaches us that "God's creation is good." The whole Bible, in their view, instructs humans "to care for and tend the creation for God."[31]

Calvin Redekop insists that "this earth is not irrelevant but is the context in which the faithfulness to God is expressed ... Anabaptist-Mennonite theology has implicitly assumed that the earth is the place where God's will is being done, as in heaven." Redekop believes that "even for a dualistic theology there is a God-creation relationship in which the human being may not be the central figure." God created the Garden (the environment) first, which provided "all the necessary elements for human survival ... Thus God's purpose for humanity must include a creation." Redekop explains that "the Genesis story considers the nonhuman world worthy of preservation and nurture," proving "that God claims for his own the nonhuman world as well as the human, and we are but temporary caretakers of the fantastically complex system we call the earth."[32]

The relation of creation ethics to the fall of humanity described in Genesis argues against the Christian theological tradition of anthropocentrism, according to Redekop. He rejects traditional interpretations that condemn the fallen world as wholly evil, since "what God has created and loved cannot be fully evil . . . The goodness of the original creation is proclaimed in the creation story, and it is difficult to propose that the Fall makes the creation totally evil or totally fallen."[33]

Comstock's interpretation of the Genesis story also embraces a nonanthropocentric creation theology, but it has a different focus— the promotion of biblical vegetarianism. Citing Gen. 1:29–30, he claims, "Before God made humans, God made animals . . . God's first words to humans" concerned "our relationship to animals" and instructed "us to eat *only* plants."[34] Finger, too, explicitly rejects anthropocentrism, particularly in his critiques of process and feminist environmental theologies.

Contemporary Anabaptist thinkers' reinterpretation of the fall even suggests an Anabaptist environmental theology of salvation, including a green Christology. Meyer, for example, defines sin as ecological disaster and redemption as the renewal of creation. "Through the disobedience of people to basic spiritual and natural laws, God's good earth has been seriously degraded." In Redekop's as in Meyer's economy of salvation, sin is rejection of God as Creator and the destruction of *shalom*. "To destroy a part of 'all things' is to attack Christ." This interpretation of the incarnation, "the epitome of God's working in [human] history" through human instrumentality, means that "human beings follow Christ and share with him in doing his will on earth, which includes nonhuman nature."[35]

Comstock notes that "God put animals at the Incarnation and at every other major event in salvation history," further arguing for a nonanthropocentric Christology. Redekop views ecological wholeness as an example of God's will expressed through both the creation and the incarnation. "*Shalom* is the metaphor of a kingdom, where the King is concerned about each member and also concerned about the land, which nurtures the people . . . It is a symbiotic kingdom, where the Lord, his people, the living systems and the sustaining inorganic substances are mutually serving each other."[36]

Finger's environmental Christology embraces these themes. He notes that Anabaptist Christology has historically been holistic, never separating Jesus' life from his death, the historic person from the cosmic Christ, the cross from the resurrection. Further, Finger argues that Christ functions as a member of the Godhead community in God's triune nature. For him, Christ's incarnation and resurrection in the material world demonstrate the Godhead community's significant past and present relationship with the physical realm of the created order. He proposes that the promise implicit in Christ's resurrection is the bodily resurrection of God's people, continuing God's relation to the redeemed physical world in everlasting terms.

Anabaptist theologians have maintained that God's promise of universal salvation means that God will restore *shalom* to the whole created order. Dorothy Jean Weaver, for example, has constructed a Scripture-based environmental theology that emphasizes Christ's role in redeeming all of creation. She interprets Christ's incarnation as "an upward motion in which creation is filled with and bears witness to the divine," as well as the more traditional understanding of God's downward motion toward creation. Based on his reading of Hosea 2:18–20, Meyer also insists that "all creation is included in the hope for redemption."[37] Further, Comstock interprets Isaiah 11:5–9 as promoting ecological nonviolence and pacifistic vegetarianism: "When the Lord comes in power . . . and the knowledge of the Lord fills the land, . . . all animals will stop doing harm to each other, and 'the lion shall eat straw like cattle.'"[38]

Anabaptist scholar Walter Klaassen has echoed the inductive logic of the 1766 Ris Confession: "The fullness of God's Kingdom encompasses all of creation: Whenever we see a stone, tree, animal or person we see the speech of God." Redekop ties these themes together. He proposes that reconciliation between God and humans necessarily involves creation, since humanity is absolutely dependent on the environment God created. Redekop writes that "a plan of *shalom* for humanity excluding nature would be unthinkable . . . When the creation is redeemed, it will happen simultaneously with the total redemption of humanity and nature (Rom. 8)."[39]

Indeed, Anabaptist environmental theologians seem to suggest that redemption is a process of reconciliation actively undertaken not only by Christ but also by human beings. "Mennonite theology and ethics historically have focused on the reconciliation of God with a people who were estranged from him," explains Redekop, but "salvation for humanity will come [only] when people are able to live in obedience to God, love their fellow humans and be able to live in harmony with the created earth."[40]

A biblical environmental ethic based on Romans 8 (specifically verse 21) suggests a restoring of the creation: "Christians are obligated to begin the restoration of the natural world now." As Dale Brown explains, Anabaptist ethics are based on the concept of discipleship. "In rejoicing about what Jesus had done for us, we are to be concerned with how we can be faithful to Jesus."[41]

This understanding of salvation and discipleship includes our behavior toward God's creation and is the foundation for an Anabaptist creation ethic, as Finger has outlined. The relationship between redemption and discipleship is central to Finger's Anabaptist theology of creation. God dwells and suffers redemptively in the world and in its creatures, and they live and move and have their being in God—not by nature (panentheism)—but by grace. Thus, as Finger notes, through God's presence the New Heaven and New Earth preserve and transform the present created order, and any work we do to sustain it becomes part of *his* work in grace.

AN ANABAPTIST CREATION ETHIC

Although most contemporary scholars have found little evidence of an early Anabaptist creation ethic, the Anabaptists in Zurich and Bern believed first and foremost that the earth is the Lord's. Anabaptist ethics are strongly identified with the values of simplicity, nonresistance and nonviolence, and community living. These ecologically sound concepts are not only good Christian discipleship, but also are three of the ten key values of the international political movement known as the Green Party. Recognition comes unexpectedly: An Ohio Amish farmer's journal was published recently by North Point Press with a preface by internationally known environmentalist poet Wendell Berry.[42]

The Anabaptist ethic of simplicity has been applied to environmental concerns, most notably by Doris Longacre and Delores Histand Friesen in the *Living More with Less* book series. According to Friesen, "recycling, conserving, [and] sharing are the firstfruits of the harvest of justice."[43] Friesen's daughters Rachel and Ingrid co-wrote, with Monica Honn and Noreen Gingrich, new words of environmental commitment to the tune of *Jesus Loves Me:*

Let's take care of God's good earth,
water, forests, air, and soil.
Don't toss out that used tinfoil.
Ride your bike and don't burn oil.
Love one another,
share with each other,
save this great earth of ours,
and learn to do with less. . . .
Learn to enjoy the simple things. . . .
Take care where you spend your cash.
Wear used clothing, mend your rips.[44]

These words, written by Mennonite children, are perhaps one of the clearest statements of an Anabaptist environmental ethic yet recorded.

The Anabaptist theology of nonresistance and nonviolence has also been applied to environmental issues. Gayle Gerber Koontz has stated that nonviolence is so central to Anabaptist ethical thought that it is the foundation for Anabaptist dialogue with other faiths. Embodying peace and a "gospel of reconciliation" means being "respectful and defenseless" in our relationships. John Ruth has called nonresistance the most intrinsic thought in the Mennonite attitude. Though it "once meant staying out of the world's power struggles," and still does to many, some Mennonites and Brethren have become actively involved in challenging unjust social structures, including the destruction of the environment.[45]

Early Anabaptist leader Menno Simons, citing Paul, explained the relationship between Christian pacifism and Christian activism: "We wrestle not against flesh and blood, but against principalities, against powers, against the rulers of the darkness of this world, against spiritual wickedness in high places. Wherefore take unto you the whole armor of God, that ye may be able to withstand in the evil day, and having done all, to stand. Stand, therefore, having your loins girt about the truth, and having on the breastplate of righteousness; and your feet shod with the preparation of the gospel of peace."[46]

This biblical mandate illuminates a model for nonviolent Anabaptist ecological activism. Meyer applies this ethic of discipleship and nonviolence to ecology. Modern nuclear, chemical, and biological weapons are ecologically destructive, while human greed has so stressed the natural environment through overpopulation, pollution, overconsumption, deforestation, and soil erosion that eruptions of violence and war become inevitable. Meyers's response: "For Christians who are witnesses to the gospel of peace there is really no choice—there must be a faith concern about the environment."[47]

Finger and Redekop have focused most on this issue, outlining an Anabaptist creation ethic in which the central issue is "human ruthlessness toward creation, which could now result in total destruction for both humanity and the created order." According to Redekop, Anabaptist ethics must include a cry for the "halt to the victimization of nature by whatever manner, for we now see more urgently that all things move and have their being in Christ."[48] Thus, in Anabaptist environmental ethics, ecocide is placed in the context of Christology.

Redekop equates the fall with the beginning of human violence and violence with the use of coercion in relationships and desecration of the created order, as well as of human beings. "An ethic derived from a theology of the original goodness of God's creation and the interdependence of all things forces us to expand the ethic of nonresistance . . . from the community of faith . . . to the larger ecological community." The definition of nonresistance thus grows to embrace "respect for everything God has created." The "way of peace, nonresistance, *shalom* . . . cannot restrict itself to peace among people, for that is not only theologically false but also practically impossible. We cannot do violence to our kin without implicating nonhuman nature. And . . . violence to nonhuman nature means disobeying God's will in how we relate to one another." Comstock agrees, but for him, truly Anabaptist pacifism means vegetarianism. "We humans, fallen and bloodthirsty as we are, retain the capacity to choose not to kill."[49]

The Anabaptist community also addresses environmental concerns. In Finger's creation theology, even God's very nature is communal, with each of the three persons of the Godhead in continuous mutual dynamic agapeic interrelation. Anabaptist community is expressed through mutual aid and the sharing of material resources. Anabaptist groups work cooperatively on peace and other voluntary service concerns through such pooling of goods. Mutual aid means that the basic welfare of each member of the group comes before the satisfaction of personal interests.[50]

Historically, the congregation is the highest human biblical and ecclesiastical authority. Even today, "Scripture is not to be interpreted so much by the individual as by the body of believers . . . Anabaptists stress that answers to questions about . . . Christians' social responsibility must be based on biblical truth that is tested in congregational meetings and regional conferences."[51]

The tendency to put the good of the group before personal interests also acts as a check and balance to American individualism, a system that drives consumerism and other ecologically thoughtless behaviors. In contrast, the Anabaptist vision "is a web of relationships, of bonding, of connectedness. It may be best encapsulated not in precise definition, but in poetry and hymns, in singing and in story, the way of the Psalms," observes Robert Kreider.[52]

Comstock emphasizes the mutual accountability aspect of Anabaptist community, wherein Anabaptist sisters and brothers challenge each other "to live more simply and peacefully."[53] Amish church members, for example, must "live within the boundaries set for their lives by the community." Though "great variety exists from community to community, . . . within a given district, one will find great uniformity."[54] Church councils even establish farming rules, including strict limitation of the use of pesticides and pollution-causing farm machinery. Surplus from the farms is canned, pickled, and given away to those in need. Horse-based, rather than tractor-based, farming is not only less polluting, but also "helps build community . . . Among the Amish the value of community controls technology, not the other way around," contends Levi Miller.[55] From this Anabaptist perspective, community living and environmentally low-impact living intertwine.

THE FAITHFUL, PRACTICAL ANABAPTIST RESPONSE REGARDING THE CREATED WORLD

Given the Anabaptist understanding of discipleship as practical religion, an analysis of the historical record of Anabaptist practice is in order. "Mennonites and other Anabaptist groups have usually stressed *living* the faith," explains Marlin Miller, "giving more weight to Christian practice than to standardized doctrinal formulations." All Anabaptists share the belief that faith "in Jesus as the Son of God and Savior can never be separated from following him in everyday behavior" and that the Scriptures "provide the primary standard for faith and life," says Marlin Miller. From the earliest days of the Anabaptist movement, Miller says, "Anabaptists have sought earnestly to be a biblical people ... in practical terms, ... placing the emphasis on *applying* the message of Scripture."[56]

Menno Simons, again citing Paul, described this relation between Christian thought and action: "A man's walk, word, and visage testify concerning a man, and the thoughts of his heart also testify what he is." Alexander Mack, founder of the Brethren, also stressed the importance of faithful Christian action: "Though we are saved by faith in Christ alone and not our simple works, saving faith will mean that we will do what Jesus wants."[57]

The Anabaptist belief that ethics is applied theology is itself a resource for environmental ethics. No abstract ethical theory will suffice in the current ecocidal context. Any environmental ethic must be lived to be effective. As Levi Miller asserts, "our view of Christian discipleship makes us skeptical of someone who talks too much about beliefs apart from culture, practice, structures, and issues of daily living."[58] This Anabaptist cynicism toward dogmatic theology is a healthy check against green theology in the abstract.

The Anabaptist practice of simple living—the practical counterpart of the ethic of simplicity—is an environmentally sustainable historical tradition based on Anabaptist interpretation of Scripture. Anabaptists believe that "living simply and sharing the world's resources" are the heart of a kingdom lifestyle, Christian discipleship as taught in the Bible.[59] As David Kline has shown, Amish Anabaptists, in particular, "value the simple things—religion, family, land and animals." In 1990, at least 130,000 of the half million Anabaptists in North America still dressed plain, and the Amish population has continued to grow since then. "Where the world cries, 'More, more!' the Amish give thanks for 'enough' ... The Amish believe their way of life represents a viable alternative to the modern way."[60]

In terms of ecological sustainability, they are right. "The Amish do not use modern equipment such as tractors and combines for their farming" because of religious beliefs "determined by community process, past and present, and backed up by economic realities."[61] Home-centered family life binds the community even more closely together, as well as limiting the use of polluting transportation technology. "Hand-built Amish buggies are ecologically sustainable alternatives to automobiles not only because they avoid air pollution and fossil fuel consumption but also because they do not need to be scrapped, repaired, or replaced with such frequency."[62] The horse-drawn equivalent of a pick-up truck, the buggy results in considerable savings in fossil fuel and emissions.[63]

Comstock asserts that traditional Anabaptist "mixed family farms are the most politically viable institution for meeting our moral obligations concerning food production, rural economies and future generations." Menno Simons himself exhorted clergy to simplicity and agricultural life: "Rent a farm, milk cows, learn a trade if possible, do manual labor as did Paul, and all that which you then fall short of will doubtlessly be given and provided you by pious brethren by the grace of God, not in superfluity, but as necessity requires."[64]

Even liberal Protestant theologian Reinhold Niebuhr recognized the spiritual value of Anabaptist agricultural life. He called the proverbial Mennonite farm "a Protestant surrogate for the monastery." Historically "wedded to the land," many Anabaptists still farm. "A feeling for good land is in their history," Ruth observes. For two centuries, Dutch Mennonites thrived "as premier agriculturalists of Russia. Their descendants now farm land stretching from Oklahoma to British Columbia."[65]

Family farming is an Anabaptist tradition in which Christian stewardship of land and animals is, for the most part, humanely and sustainably practiced. God asked "the earliest man and woman to name the animals and cultivate the earth," laying the biblical foundation for Anabaptist stewardship of animals and land.[66] Anabaptist stewardship has even been a source of Christian witness and conversion. Ruth reports that "5,000 Massai ex-warriors in East Africa," affected by Mennonite service activity, "recently said they would like to be Mennonites, because Mennonites understand cows."[67]

Amish and Mennonite closeness to the land leads to a theology that arises from daily life practice, a reversal of the relationship most Christian theologians posit between theology and ethics. Amish farmer and writer David Kline proposes that, "if one's livelihood comes from out of the earth, from creation, on a sensible scale, where we are part of the unfolding of the seasons, experience the blessings of drought-ending rains, and see God's hand in all creation, a theology for living should be as nat- ural as the rainbow following a summer storm." Amish farmer Mose Kaufman agrees: "Living on a farm, planting and reaping, makes you conscious that all our sustenance comes from an all-wise, all-powerful creator . . . Having in mind where it all comes from makes you feel hum- ble and thankful for the abundance of nature."[68]

Anabaptist agricultural life, then, is both a witness to and a resource for Anabaptist environmental theology. As simple agriculturalists, An- abaptists have been living and working for environmental wholeness perhaps since the beginning of the Anabaptist movement. Contempo- rary Anabaptists have recently begun to construct theology, ethics, and service projects that are ecologically sound. Anabaptist history, there- fore, both recent and remote, provides a resource for an Anabaptist en- vironmental ethic.

THE CONSTRUCTION OF A FAITHFUL ANABAPTIST GREEN THEOLOGY AND CREATION ETHIC

The task of constructing a green theology and ethic that are faith- ful to Anabaptist beliefs must begin with the acknowledgment of its in- volvement in the current ecological crisis. Ken Gex advocates ecological "kingdom living"—drastically reduced consumption of resources by mid- dle-class Americans:

> Unlike other social problems we face, this one involves each of us because we add directly to the problem and each of us can help to reduce it . . . What we purchase (demand) from "big businesses" (supply) and what we do with our waste are critical to environmental concerns . . . Although fewer of us actually work the land today, we still depend on its fruit for our well-being. As stewards of the land [Gen. 2, Rom. 8], we need to change our "demand" in the marketplace and reduce the volume of our waste.[69]

Schrock-Shenk also urges North Americans not to neglect global environmental problems: "The work of addressing root causes of natural disasters extends to North America."[70] Anabaptists must therefore con- struct a green theology mindful of North American consumption of a disproportionate share of global resources, contribution of a disproportionate share to the world's pollution, and spreading of unsustainable agricultural and consumer practices around the world. In other words, North American Anabaptist green theology must be especially humble, not paternalistically self-righteous but self-consciously repentant.

An Anabaptist environmental theology will have its theological roots in Anabaptist biblical materialism, Christ's love ethic, the martyr ethic of witness, the community ethics of nonconformity and mutual accountability, and a theology of repentance and rebirth. James McClendon defines Anabaptist biblical materialism as a theology of "our embodied selves," equipped by the creator with physical needs. The challenge of meeting these God-given physical needs leads to the development of our moral faculties. "Our share in the created order is marked by our share in Christ—who on earth also possessed these native drives, natural needs, and nascent resources of the body."[71]

Genesis 2:15 is the source of this material anthropology, "an ecological understanding of our organic selfhood," intrinsically related to the "tilling and keeping of the garden" (nature). The life of each Christian is organically related to the life of the earth itself. Even the Christian sacraments are intrinsically material as well as spiritual. "Baptism is the baptism of our bodies; in the Lord's supper we feed body and soul alike and at once; it is with physical, fleshly ears that we hear—the word of God."[72] Gex cites Genesis 2 as a foundation for Anabaptist biblical green theology: "In Genesis 2, . . . the concept of earth stewardship is found as God instructs those created in his image to 'take care of the land.'" In Romans 8, Paul informs us that, as believers, we have a responsibility "to help liberate creation from its bondage of decay brought on by the fall of humankind."[73]

Kropf elaborates Anabaptist biblical materialism from an ecological perspective: "Throughout the Psalms, mountains and hills, birds and creeping things, wild and tame animals and all people are called to bless God's name forever . . . Every creature joins a vast song of praise in God's honor . . . In response to [God's] lavish creativity and generosity, human beings respond by adoringly blessing God, gratefully recognizing 'the immensity of God's goodness.'" McClendon cites the Psalms, proclaiming that God "is the absolute context, the everlasting environment (Psalm 139) of life."[74] Whatever else it is, Anabaptist environ-

mental theology will always be Scripture-based, even in its understanding of material existence.

The Anabaptist love ethic is derived from this biblical theology, as
is the Anabaptist ethic of nonresistance. Both are central to Anabaptist
faith and share "a high view of the sacredness of life."[75] According to
Menno Simons, "love is the total content of Scripture." The "sincere and
unfeigned love of one's neighbor" is the sign by which the church of
Christ (as opposed to the false church of the antichrist) is known. "By
this shall all men know that ye are my disciples, if ye love one another.
John 13:35 ... Wherever sincere, brotherly love is found without
hypocrisy, with its fruits, there we find the church of Christ."[76]

The love ethic, recognized by its fruits in daily practice, is central to
any Anabaptist ethics and will be central to an Anabaptist environmental ethic as well. Jim Rich applies this lifestyle to the earth itself: "We
must fall into the agony of unconditional love as one does for a cherished
friend who is dying in the prime of life—a love willing even to give up
life if it would save the friend ... All of us on earth can live well, though
in a radically simplified manner, on the earth's bounty. Sacrifice is not
the issue. The issue is love. This is a spiritual matter."[77]

Rich has extended this respect beyond the human neighbor to the
whole created order. "Providing more of our [own] energy and food
needs ... strengthens our love bond with the earth ... To reconnect with
the earth ... is fundamentally a heart connection. The spiritual aspect"
of "hands-on labor" is that it is "powered by love."[78] Thus, in an environmental Anabaptist ethic, the Anabaptist theological traditions of bib-
licism, the love ethic, and simplicity are all linked and interwoven with
deep ecology.

The Anabaptist understanding of witness must also be a fundamental part of any Anabaptist environmental attitude. As William Klassen
proclaimed at the 1994 triennial peace colloquium, "Our mandate is to
witness, not to convert." In terms of an ecologically sustainable lifestyle,
Anabaptist witness means "a return to more-with-less living," or simplicity. More-with-less living, most notably outlined by Longacre and
Friesen, entails sacrifice by contemporary Anabaptists who have assimilated to North American materialism, consumerism, and prosperity.
"Following the pattern of behavior Christ taught and lived may ... be
costly," Ruth warns, but "emphasis on the cost of discipleship goes back

to the Anabaptists' earliest beginnings."[79] Though a daunting task, true discipleship—Christlike simplicity and selflessness—is the witness an authentically Anabaptist environmental theology must proclaim.

Nonconformity is another hallmark of any truly Anabaptist ecological ethic. According to Marlin Miller, Anabaptist ecclesiology understands the church as the community of believers called by God out of an unbelieving world, "the visible body of those who have voluntarily confessed their faith in Jesus Christ and have committed themselves to follow him in life." In the earliest days of the Anabaptist movement, nonconformity was political as well as spiritual. Anabaptists were outlaws, since "rejecting infant baptism in favor of believers' baptism amounted to an act of civil disobedience." Like the early Anabaptists, contemporary believers must apply the standards of the Sermon on the Mount to "Christian conduct here and now." Miller warns, however, that since Jesus' teachings "run counter to sinful human tendencies and to the structures of society in general, . . . faithfulness to his way leads to conflict with social structures."[80] Since the social structures of the contemporary world are bent on the destruction of the divine creation that sustains us, we will have to counter and oppose them. With our Amish sisters and brothers, *all* Anabaptists will have to self-critically ask ourselves the "central question . . . , 'How can we live together as God's humble, obedient people *in* the world but not *of* it?'"[81]

Though Anabaptists or others may not choose to live an Amish lifestyle, we can still embrace the concept of nonconformity. Such nonconformity will be vital to a truly environmental Anabaptist ethic, for we cannot conform to the consumeristic exploitation and pollution of the natural world if we, and especially future generations, are to survive. Some contemporary Anabaptists are modeling new ways of nonconformity and civil disobedience to environmentally destructive social structures: "Grandchildren of . . . Mennonite farmers may be found today climbing over fences around nuclear installations."[82]

Dale Brown reflects that "we serve this age as pilgrims, strangers, aliens, and sojourners because of a sense of expectancy that the future age can break into the present. As much as possible, by grace, we are to live now as if the kingdom has already come."[83] Kingdom living, in an ecological sense, will mean nonconformity to the social structures that tempt us to indulge consumeristic, materialistic greed and selfishness.

The Anabaptist tradition of mutual accountability is one means of achieving such nonconformity. "The emphasis on mutual admonition and correction" is "traditionally . . . based on Matthew 18:15–18."[84] Mutual accountability is ideally practiced in Anabaptist congregations in the following way: "A small group of members . . . work with the pastor and the persons concerned to provide discernment, counsel, and support to both the individuals and the congregation." For most "contemporary North American Mennonites" accountability leads to "restoration . . . not to breaking off all relations." In other words, church members who succumb to temptations of various kinds receive the kind of admonition and support that encourages repentance, forgiveness, and restoration to the community. "Most [Anabaptist groups] have made the peace position a matter of mutual admonition and correction."[85] This ethic of mutual accountability and process of reconciliation must also be central to any Anabaptist ecological ethic. As Anabaptists make the environmental position a matter of mutual admonition, we can challenge and support each other to live in ever more sustainable ways for the sake of God's creation and future generations.

As mentioned earlier, earnest repentance must also be included in an authentic Anabaptist environmental lifestyle. Menno Simons cautioned, "Both openly and privately, and studiously avoid sin." Further, he warned, "If not sincerely repented of, [sin's] outcome will be eternal death." This is certainly true with regard to sins of wanton destruction of God's creation. Menno constructs an outline of the fruits of repentance to help with this task: "It is not always repentance when men say, I have sinned . . . Repentance is a converted, changed, pious, and new heart, a broken and contrite, sad and sorrowful spirit, from which come the sorrowful tear and lamenting mouth, a genuine forsaking of the evil in which we were held, an earnest and hearty hatred of sin, and an unblamable pious Christian life."[86] Using Menno's guidelines as a measure of true repentance, we can call ourselves and each other forward from our current, inadequate environmental efforts. Those of us who have begun to attempt an Anabaptist environmental theology have barely begun to acknowledge that we have sinned.

Truly sorrowful contrition requires that we know and believe in the reality of ecological crisis and that we fully accept our share of responsibility for it. Knowledge of sin and admission of guilt are still one step removed from true repentance—actual change of behavior. A practical conversion, from environmentally destructive behavior to ecologically sustainable behavior, must be the goal of Anabaptist environmental ethics.

Menno Simons acknowledges the difficulty of our task, offering a prayer for rebirth when not "regenerated" after baptism as we should be. Menno reminds us of God's promise in baptism, resurrection, and re-demption that "all things are become new." Yet our renewal (as good and faithful stewards of God's creation, for example) is ultimately dependent on God's grace, not on our own self-righteous efforts. "Lord, send forth Thy Spirit, and we will be created, and Thou wilt renew the face of the earth."[87]

CHAPTER 13

The Environmental Challenge before Us

Calvin Redekop

Many people feel that it may be too late to save our world and that the environmental challenge is too large a task for humans to contemplate or understand, much less solve. The question as to whether human beings can actually take a self-conscious, objective, and constructive stance toward the total environment, which largely creates human nature and consciousness in the first place, is daunting. As René Dubos suggests, "The distinction between organism and environment becomes quite blurred . . . when one considers biological nature and external world not as separate static entities, but as interacting components in complex dynamic systems."[1] That being the case, can humanity be expected to become critically self-conscious about itself and its environment?

This book begins and ends with the well-accepted axiom that humans profoundly influence the environment: "All organisms impose on their environment characteristics that reflect their own biological and social nature."[2] We further assume that philosophical and religious systems do influence the basic relationships of humans to the environment. The important question, therefore, remains: What can and should humans do about their relationship to the environment?

In the final chapter I will explicitly state what has been implicit throughout the book: The environmental crisis has developed because of human hubris. Humans have assumed throughout history that by their rational powers they can know the nature and purpose of human existence apart from their relationship to the created world. One of the many insights given us by Native Americans is that we have not listened to nature and, consequently, have not recognized or admitted our interdependence with nature. The Western world's relationship to the nurturing creation, recently exacerbated by modern science, has been basically a one-way conversation.[3]

Religion, specifically Christianity in the West, has played an important role in this one-way relationship. This proposition is illustrated by the great debate created by Lynn White's claim that "Christianity bears a huge burden of guilt for the ecological crisis." A great deal of research and analysis was unleashed, including the report that "the sociological survey literature that exists on the subject seems to agree that the more 'Christian' or biblically oriented one is, the less one is concerned about the environment."[4] Christians generally did not want to admit that their relationship had been one way—that of domination of the earth.

The general view is that Christianity has had a major influence on the environment and that, on balance, that influence has been far from positive. What has been lacking in the Christian community is humility—a candid and open discussion of the positive as well as negative aspects of human interaction with nature. Many Christians challenged and rejected White's claim of Christianity's negative impact. In a rather evasive and self-serving manner, many Christians protested that only a "perverted" Christianity could have had a negative influence.[5]

RECENTLY DEVELOPED ENVIRONMENTAL ETHICS

Nevertheless, in the relatively short time since the environment has been a concern for the Christian community, several identifiable positions have emerged.

The Christian Stewardship Ethic

The Christian stewardship ethic begins with the biblical command giving humans dominion over the earth. A positive cast would "interpret [dominion] as a divine charge to be good stewards and to take care and protect the Creator's creation."[6] When dominion is rightly interpreted, all is well. "The ecological crisis thus arises from human sinfulness by disobeying this commandment"; to "be Christian is to be ecologist." The more conservative and fundamentalist wings of evangelicalism consider even this suspect and view all "environmentalism as paganism."[7]

Nevertheless, this position makes a contribution to the environmental crisis. A protective and conservationist stance toward the natural world slows down its degradation but does not offer a significant solution to the fundamental causes of environmental damage.

The Ecojustice Ethic

Supported by mainline Protestantism, the ecojustice ethic "focuses on linking environmental concerns with church perspectives on justice issues such as the just sharing of limited resources and the real cost of environmental problems. It thus combines an already present Christian social justice framework with environmental concerns."[8] This orientation focuses on changing institutional structures by working with political, economic, and social processes to achieve justice and environmental preservation.

Identifying the structural and institutional injustices resulting in environmental destruction and working to change them are necessary and positive steps in environmental redemption. The challenge in this position is to achieve sufficient consensus among groups of people to form a unified and coherent plan of attack.

The Wise Use Ethic

This ethic, held by some conservative evangelicals, suggests that "God is indifferent to . . . Creation, and thus it has no moral status. Furthermore, the best strategy for achieving the welfare of present and future generations is not conservation, but rather economic growth and 'resource substitution.'" These people hold that "there is absolutely no concern for Creation. God made it for us, and we can do to it whatever increases human well being."[9] This position is so foreign to our position that it will receive no further comment.

The Creation Spirituality Ethic

The creation spirituality ethic also starts with the Genesis story but, transformed by modern science, the story becomes universal. The privileged status of humans is removed, and they become a part of the whole. "The chief obstacle to an ecological work is not human sin or injustice, but overcoming the dualism of the western world-view so that we can see the creation as a whole."[10] The new physics and old medieval mysticism together help overcome traditional dualisms. Creation spirituality can use religion to understand the significance of the ongoing revelations of science and can apply them to re-create a more unified relationship of the human and natural world.

This focus can be of great significance, but any form of spirituality, especially in regard to the environment, must face the challenge of "spiritualizing" the struggle for ecojustice and making it an intensely individualistic and subjective practice. Saving our natural habitat will require collective, political, and concrete planning and action.

Ecofeminism

Ecofeminism is a recent philosophical and social movement that spills over beyond Christianity; it emerged as a definable entity in the midseventies with the publication of Rosemary Ruether's *New Women / New Earth* in 1975. Adams maintains that "ecofeminism argues that the connections between the oppression of women and the rest of nature must be recognized to understand adequately both oppressions." Ecofeminists "insist that the sort of logic of domination used to justify the domination of humans by gender, racial or ethnic, or class status is also used to justify the domination of nature."[11]

There is, however, no unified position in ecofeminism. Karen Warren states that, "just as there is not one feminism, there is not one ecofeminism." And because ecofeminism is still so new, its position and influence are still evolving. According to Warren, there are several minimal claims to which all adhere: "There are important connections between the domination of women and the domination of nature and understanding the nature of these connections is necessary to an adequate feminism, environmentalism or feminist or environmental philosophy."[12] Unquestionably, the basic philosophical orientation of ecofeminism, whether Christian or not, is irrefutable. The survival of the earth and humankind hinges on exorcizing the oppression and domination of any sector of humanity or nature.

The Anabaptist Environmental Ethic

The Anabaptist environmental ethic, as interpreted in this book, is a derivative of the Christian heritage.[13] But it has long differentiated itself from the mainline denominations, for, as the chapters by Hiebert, Weaver, Klaassen, Finger, Hart, and Ackley Bean suggest, the Anabaptist perspective proposes a "third way," neither Catholic nor Protestant. Briefly summarized, this heritage has always believed that God has been concerned with the formation of a "People" who would fulfill God's wishes *on earth*. This concept has been interpreted in the Anabaptist tradition as the rule of God in human society—"the kingdom of God."[14]

According to this view, "not only is creation an expression of God's rule, but continuing order in the natural and human worlds is also based on God's kingship. The kingdom of God means a redeemed earth; earth is the scene and object of the kingdom of God." The "new age" toward which Christians strive is "the restoration of the entire cosmos to its original state; that is, the renewal of creation. In the end, there will be no more separation between the earthly and heavenly realms."[15]

The various dualisms that have tended to separate the material from the spiritual or heavenly throughout history have thus been rejected by the Anabaptist world-view; the issue was not a spiritual/material dualism but a dualism of kingdoms, which begins with separating the obedient, redeemed community from the disobedient and rebellious society. Its missionary goal is bringing all of God's children and creation under his rule in his kingdom. This means that the creation will be a part of the redemption in concrete union of the spiritual and physical realms, as the authors and others cited in this book suggest.

CHRISTIANITY'S CHALLENGE TO CLEAN UP ITS ACT

Not all of the aspects of religion that are destructive for the environment are addressed by the six orientations discussed above, nor could they be. Further, all of humanity, whether religious or not, from early times has been involved in the processes that have destroyed our world. These consequences are more complex than what one religious movement or group could accomplish. The economic institutions and processes, science and technology, population growth, as well as the various cultural forms that have pitted one group against another wherever they sojourned, have contributed to the crisis.

Nor will everyone identify with one or even a combination of the ethics discussed, but an analysis of the environmental movement will reveal that the Christian religion in all its manifestations, including Anabaptism, has reflected the six positions and is actively involved in environmental degradation as well as attempts at redeeming it. White reinforces the importance of religion in the environmental crises with his ironic dictum that "more science and more technology are not going to get us out of the present ecologic crisis until we find a new religion, or re-think our old one."[16] He proposes that religion is a key influence in any solution to the environmental crisis. We strongly agree. We do not dismiss the vast secularization that has infused Western culture, with its negative effect on creation, but we are here focusing on the religious dimension and its effects.

What are the challenges confronting us as we respond to what has been presented in this book? First, for those who identify with the Anabaptist heritage and its orientation toward the environment and creation, the challenge is to reacquaint ourselves with that orientation toward the creation and then to live up to the ideals it presents. Put differently, Anabaptists face the great challenge of discerning the "laws" of the "kingdom of God" and transforming them into concrete reality—*praxis.* There has been considerable slippage and even rejection of this heritage among Anabaptists and, hence, degradation of the environment.

Second, for members of religious groups outside the Anabaptist tradition falling somewhere in the other categories mentioned above, the challenge is to critique and question their own tradition where it is wide of an authentic Christian interpretation of God's plan for the creation. If this dialogue has not progressed far, that is probably partly the result of little conscious effort on the part of any of the religious traditions, including Anabaptists, to explicate their position. We hope that this book provides some assistance in this ecumenical quest. If there are constructive insights and practices in the Anabaptist view of creation that are valid for other traditions, and vice versa, the invitation to engage in environmental praxis is extended to all.

Finally, for those who are not religious but are concerned about the environment, the challenge is to understand the religious sources of a great many of our attitudes toward the environment. With that understanding, they can then consciously challenge Christians of all persuasions in an attempt to stimulate them to rediscover God's plan for creation and to restore and preserve the planet, in all its natural richness, to the end of achieving a sustainable natural, social, and economic world.

PRACTICAL SUGGESTIONS FOR EVERYONE

The following suggestions reflect the hope all humans share, regardless of religious or secular, nonreligious traditions—to preserve the present and future home of all living things. We can still experience a conversion from human hubris regarding the existence of life on earth. This hope moves us all for, regardless of our traditional roots, we soon will all experience a conversion toward a sustainable relationship to our mother earth: What can we do to bring it about?

1. We can all consume less economic production and land itself. Not only will this result in less pollution and waste, but it will help us face our almost irrational materialism and thereby free the human spirit. Reduced consumption will also leave more of nature's largess for wild nature and for the many peoples of the world struggling for the necessities of life.

2. All of us can work to achieve a human population that will result in a harmonious balance regarding consumption of the world's limited space and natural resources, the increasing pollution and contamination of our world, and the termination of encroachment on the natural wildlife habitat. A finite world implies a finite growth trajectory. We must do battle with the reigning ideology proclaiming "growth, whether technology, population, or GDP, as the final good" (see chap. 1).[17] Practically, we will support on a global perspective all organizations and programs, private and governmental, that promote education and assistance in family planning, birth control, and the reduction of the number of children birthed.

3. We can all strive to leave this world a better place than when we arrived here by working to achieve an infinitely *sustainable world*. This work includes local restoration projects, such as planting trees and cleaning up polluted streams,[18] and more public and structural changes, such as supporting the various organizations and movements, public and private, that fight the growth and "development" cancers exemplified in such futile exercises as building more and bigger highways at the expense of mass transit, bikeways, walkways, and so forth.

Further, we must all support such organizations as the Nature Conservancy, which is heroically buying up land to preserve the natural habitat of the myriad threatened flora and fauna. We must work for structural change through the political process, from local to national governments. The biggest challenge will be to achieve a human order that is "regenerative rather than depletive," in which "waste equals food" for every succeeding generation's biosystem.[19]

4. Finally, we can begin by changing ourselves. As Philip Sherrard says, "Our entire way of life is humanly and environmentally suicidal. Unless we change it radically, there is no way in which we can avoid catastrophe."[20] But constructive societal change comes about incrementally. Is there a way out? As individuals (prophets?) here and there give radical testimony to the reality that life is not measured by the abundance of things, but by the harmony between the spirit and the material aspects of cosmic reality to produce a sustainable world, society can and will change.

These suggestions are offered to the end that humans, whether Christian, Anabaptist, or secular, can join together to finally reject the trend of human history described by René Dubos: "Most societies seem willing to sacrifice environmental quality at the altar of economic wealth and political power."[21] In the second century of our era, Marcus Aurelius reasoned, "If he is a stranger to the universe who does not know what is in it, no less is he a stranger who does not know what is going on in it. He is blind who shuts the eyes of the understanding."[22]

An early Anabaptist martyr, Hans Hut, had a remarkably prescient and prophetic vision of the role of creation (1527):

> The eternal power and divinity will be perceived when a man truly recognizes it in the creatures or works from the creation of the world. If God is

to make use of us or enjoy us, we must first be justified by him, and cleansed within and without: inwardly from desire and lust, outwardly from all improper behavior and misuse of creatures. When the gospel in all creatures is thus preached according to the commandment of the Lord, and man is brought thereby to understand that reason in a natural and real way is included in his own works which he does over and in the creatures, in which he acknowledges God's will towards him. For the whole world with all creatures is a book written from nature by the Spirit of God.[23]

The ultimate law of ecology (with variations found in all religious traditions, including Christianity) is that every thing is connected to all other things. This law will ultimately become universally accepted. But the future can hardly wait. During my childhood in Montana, my mother baked bread every Saturday with flour produced by the Occident Milling Company of Minneapolis; I still remember the very effective marketing slogan on each sack—"Eventually, why not now?"[24] Indeed, why not cooperate *now* with the ultimate law of ecology, since we will *eventually have to;* if we do not accept the supreme "law of interdependence" soon, our global home will become inhospitable to all living things.

$\mathcal{A}$ $\mathcal{L}$etter to $\mathcal{C}$ongress

Mennonite Central Committee U.S.
Washington Office
(202) 544–6564
October 5, 1995

U.S. House of Representatives
Washington, D.C. 20515

Dear Representative:

Mennonite Central Committee is committed to preserving and safeguarding God's creation as an integral part of its relief, service and community development work. The current Young-Pombo Endangered Species Act (ESA) reauthorization bill H.R. 2275 hampers these efforts, and betrays our role as a good steward of God's created earth.

Issues of concern in this bill include abandoning species recovery goals, removing habitat protection, and "takings" provisions alterations—all of which undermine the ability of the ESA to protect species. The ESA was initiated to prevent the destruction of animals, plants and their habitats. To weaken legislation which preserves genetic resources is to weaken one of the government's more responsible actions to mitigate human abuses of creation.

Indeed, as the Mennonite Central Committee's 1994 statement, Stewards in God's Creation, states:

The land, the rivers, and the skies cry out for healing. They "groan" for reconciliation with the human family they wish to sustain. As followers of Jesus Christ, we need to hear this cry. We need to hear and obey the command of our

Creator who instructed us to be caretakers of and at peace with the creation. Faithful stewards of the gospel are also faithful stewards of God's creation.

Stewards in God's Creation
Mennonite Central Committee
September 1994

We encourage you to recommit government to preserving the beauty and diversity of creation, and oppose H.R. 2275.

Sincerely,

J. Daryl Byler
Director, Mennonite Central Committee, Washington Office

Stewards in God's Creation

Mennonite Central Committee Statement on the Environment

The earth is the Lord's, and everything in it,
the world, and all who live in it.
Psalms 24:1 NIV

INTRODUCTION

God's purpose in Christ is to heal and bring to wholeness not only persons but the entire created order. "For God was pleased to have all his fullness dwell in him, and through him to reconcile to himself all things, whether things on earth or things in heaven, by making peace through his blood shed on the cross" (Col. 1: 19–20). MCC seeks to follow God's command in the process of incarnating God's purposes of love for all creation. This love aids the essential needs of people, including not only food, clothing, shelter and jobs, but also hope for a new life. The challenge is to help bring wholeness without jeopardizing the ability of tomorrow's people to meet their own and the needs of other creatures with whom we share God's creation.

All over God's earth, *humanity* is rendering fertile land barren. This environment which sustains us all is suffering more and more serious degradation at human hands through: destruction of productive farm and range lands; contamination of soil, water, atmosphere, and food; destruction of genetic resources, i.e. animal and plant habitats; climatic change caused by excessive use of fossil fuels and fluorocarbons; destruction of tropical rain forests; emission of sulfur and nitrogen oxides causing acid rain; more conflict over waning resources by increasing numbers of people.

The land, the rivers, and the skies cry out for healing. They "groan" for reconciliation with the human family they wish to sustain. As followers of Jesus Christ, we need to hear this cry. We need to hear and obey the command of our Creator who instructed and led us to be caretakers of and at peace with the creation. Faithful stewards of the gospel are also faithful stewards of God's creation. There have been and there are successful efforts underway within God's creation where responsible people are taking action to correct the abuses. But much more remains to be done.

BELIEF STATEMENT

As stated in the "Confession of Faith in Mennonite Perspective," Draft October 1993

We believe that God has created "the heavens and the earth" and all that is in them, and that God preserves and renews what has been created in accord with the divine will. We believe that God has begun the new creation in Jesus Christ and sustains it through the power of the Holy Spirit. We look forward to the redemption of creation and the coming of a new heaven and a new earth, where God's purposes for all creation will be fully realized.

We believe that human beings have been created good and have been called to glorify God, to live in peace with each other, and to watch over the rest of creation. We gratefully acknowledge that God has created human beings in the divine image and has given the entire human family a special dignity among all the works of creation.

Repentance: MCC acknowledges patterns of working which have contributed to some of the destruction mentioned above. We repent of the ways we have caused pollution or destruction of our Creator's work. We want to be more fully aware of the impact our lifestyle has on the global environment, and on our sisters and brothers worldwide who share God's good earth with us.

Poverty and the Environment: Poverty and environmental destruction often fuel each other. For too many, today's survival decisions destroy land, water, and stratosphere over the long term. Conversely, some impoverished communities have learned to balance their demands and use of the environment in sustainable ways. Whichever the case, MCC is committed to work in partnership with the poor to protect and to heal our God-entrusted environment.

Wealth and the Environment: Modern "prosperity" and environmental destruction often fuel each other. People having access to billions of dollars in credit and modern machinery can destroy large areas of land or pollute the atmosphere. Individuals living in a consumptive culture also make choices that damage the environment and consume more than their fair share of the world's resources.

Commitments of MCC: Safeguarding God's creation is not an "add on" to MCC's commitments. Preserving the creation must be integral to who we are and what we believe. Since 1920, MCC has been concerned with good soil and its capacity to produce food. Our intent has been to preserve God's creation "in the name of Christ." We have engaged in reforestation, water catchment, soil conservation, erosion control, etc. The printing of materials, e.g., *More with Less Cookbook* and *Living More with Less*, illustrates MCC's environmentally friendly efforts. And MCC has persisted in peace work which reduces destruction of God's good creation.

In its attempts to be environmentally responsible and faithful, MCC will give primary concern to the following areas of activity:

1. Become more aware of and resist wherever possible the evil of environmental degradation.
2. Inform and educate in the mission of stewarding God's creation.
3. Live faithfully as stewards, protectors and reconcilers of God's created earth.
4. Conserve energy by recycling, car-pooling, avoiding over-heating and cooling in our offices and homes, etc.
5. Work with the poor in their efforts to protect the environment while also feeding them.
6. Support and cooperate (with private and public organizations) any actions which advance environmental improvement. Conversely, oppose legislation and actions which endanger the physical environment.
7. Support and plan programs that assure protection or restoration of God's creation by: analyzing potential positive/negative impacts on the environment; designating a staff person to monitor these commitments and provide feedback as appropriate; providing program to administrators with questions and guidelines which become a part of the annual planning/reporting cycle.

Planners are requested to respond to the following questions and tasks to the degree that they are relevant and helpful, acknowledging that each program and region encounters differing environmental questions. Additional screening resources are available from area administrators.

1. What are the issues in your region?
2. How do the activities of program responsibility positively or negatively impact the environment?
3. Can negative impact be changed? How? What are the risks involved? What benefits can accrue?
4. Begin to collect and record relevant data which help us to be more specific.

Approved by MCC Executive Committee, Item No. 29, September 1994

Notes

INTRODUCTION

1. The settlement consisted of Mennonites who had moved from Minnesota looking for more and cheaper land. They had found it on the Fort Peck Indian Reservation, which had been opened to white settlement in 1913 and which allowed homesteading.

2. I remember the prayers of the homesteaders, with voices raised amid tears, imploring God to send desperately needed rain. Confessions of sin and weak faith interlaced the petitions.

3. See Kittredge, "Home Landscape."

4. *Discover*, August 1995, 63.

5. And to bring the past into closer connection with the future, Diamond suggests that "our Pacific Northwest loggers are only the latest in a long line of loggers to cry, Jobs over trees" (68).

6. Some might argue that the possibility of atomic annihilation has been the greatest crisis humanity has or will ever face, making the environmental crisis a mere sideshow. We disagree. It is precisely the subtlety of the degradation of the environment, which can bring us past the point of no return before we realize it, that is the crucial issue.

7. Toynbee, *A Study of History*, 244.

8. Schnaiberg, *The Environment*, 29.

9. The Chicken Little story is probably one of the most powerful socializers into the optimistic denial stance ever devised.

10. Commoner, in Disch, *The Ecological Conscience*, 118.

11. The concept of a sustainable creation refers to a global cultural and economic system in which all wastes and by-products of human activity become food (either biological or technical nutrients) for succeeding life systems. See McDonough and Braungart, "The Next Industrial Revolution," 88.

12. It is difficult to state categorically which activities are environmentally destructive. In the case of the Mennonites, their great contribution to the draining of the swamps in the Friesland of the present Netherlands is a case in point.

13. I hope readers will not recoil at my use of the term *Great Spirit* to refer to the transcendental power or spirit that has been recognized and defined for themselves by most, if not all, societies in history, including Western Christendom.

14. Gore, *Earth in the Balance*, 259.

CHAPTER 1: ECONOMICS, DEVELOPMENT, AND CREATION

1. Hawken, *The Ecology of Commerce.*
2. World Bank, *World Development Report*, 1995.
3. Cited in Rich, *Mortgaging the Earth*, 263.
4. Levinson, "Watching the Tide Rise," 47.
5. Korten, *When Corporations Rule*, 11.
6. Hawken, *The Ecology of Commerce*, 3.
7. Brown, Flavin, and Postel, *Saving the Planet.*
8. DeWitt, *Earthwise*, 33.
9. Cobb, Halstead, and Rowe, "If the GDP Is UP?" For additional technical details of such recalculation methods, see the appendix to Daly and Cobb, *For the Common Good.*
10. Daly, quoted in Zachary, "'Green Economist' Warns." See Daly, *Beyond Growth*, for the fuller treatment of this concept.
11. The essay is reprinted in Daly and Townsend, *Valuing the Earth*, 297–309.
12. Ibid., 302.
13. The analysis is developed in the introduction to Daly and Townsend, *Valuing the Earth.*
14. Platt, "Dying Seas."
15. United Nations Environment Program estimate, reported in McNamara, *Africa's Development Crisis*, 13; Daly and Cobb, *For the Common Good*, 59.
16. Calculations by Vitiousek et al., as cited in Daly and Cobb, *For the Common Good*, 143.
17. Hawken, *The Ecology of Commerce*, 26.
18. Naess and Sessions, quoted in Daly and Cobb, *For the Common Good*, 377–378.
19. Solow, "Sustainability."
20. Daly and Cobb, *For the Common Good.*
21. World Resources Institute, *World Resources*, 8.

22. Hawken, *The Ecology of Commerce*, 21–22.

23. Ibid., 3.

24. Brown, Flavin, and Postel, *Saving the Planet*, 28–29; Hawken, *The Ecology of Commerce*, 84.

25. Rich, *Mortgaging the Earth*, 242.

26. For a fascinating account of past petroleum supply prophecies and realities, see Brown, "Outlook for Future Petroleum Supplies."

27. Daly and Cobb, *For the Common Good*, 198.

28. Hawken, *The Ecology of Commerce*, 40–41.

29. Solow, "Sustainability," 181.

30. Hawken, *The Ecology of Commerce*, 29.

31. This concept was developed by William Rees and colleagues at the University of British Columbia. See Wackernagal and Rees, *Our Ecological Footprint*.

32. World Commission on Environment, *Our Common Future*, 1–6; Wackernagel and Rees, *Our Ecological Footprint*, 97.

33. Todaro, *Economic Development*, 202.

34. Wackernagel and Rees, *Our Ecological Footprint*, 85, 15. This image is another argument for population control, this time by the developed nations.

35. McNamara, quoted in Brown, Flavin, and Postel, *Saving the Planet*, 24.

36. United Nations Development Program (UNDP), *Human Development Report*, chap. 3.

37. Whalen, *The Anxious Society*.

38. Todaro, *Economic Development*, 145.

39. UNDP, *Human Development Report*.

40. Fromm, quoted in Goulet, *Development Ethics*, 130.

41. Ibid., 135.

42. Goudzwaard and de Lange, *Beyond Poverty and Affluence*, 159.

43. Korten, *When Corporations Rule the World*, 81.

44. Goudzwaard and de Lange, *Beyond Poverty and Affluence*, 116.

45. As they flow through the economy, goods and services proceed through a series of stages, from the extraction of raw materials to the use and disposal of manufactured products.

46. Brown, Flavin, and Postel, *Saving the Planet*, 22.

47. Statistics cited in Korten, *When Corporations Rule the World*, 83, and *Co-op America Quarterly*, fall 1994, 12.

48. Postel, "Carrying Capacity," 16.

49. World Resources Institute, *World Resources*, 108, table 6.1.

50. Schmookler, *Fool's Gold*, 91.

51. Hawken, *The Ecology of Commerce*, 186.

52. Korten, *When Corporations Rule the World*, 31–32.

53. Ibid., 32.

54. Wysham, "Ten-to-One Against."

55. Strong, quoted in Serageldin, "Ethics and Spiritual Values."

56. Wackernagel and Rees, *Our Ecological Footprint*, 105, 104.

57. See, e.g., Daly and Cobb, *For the Common Good*; Hulteen and Wallis, *Who Is My Neighbor?*; Rich, *Mortgaging the Earth*; Korten, *When Corporations Rule the World*; Goulet, *Development Ethics*; and Wuthnow, *Rethinking Materialism*.

58. Solzhenitsyn, *From under the Rubble*, 105–106, 136–137.

59. Daly and Cobb, *For the Common Good*.

CHAPTER 2: SCIENCE, TECHNOLOGY, AND CREATION

1. Deevey, "The Human Population," 198.

2. Brown, *State of the World*, 4.

3. Schumacher, *Small Is Beautiful*, 140, 142, 143.

4. Ibid., 143.

5. Harden, "Nuclear Reactions," 14. See also his *A River Lost*.

6. Harden, "Nuclear Reactions," 14; Booth, "Ecosystem Paradoxically Glows."

7. Lucky, "What Technology Alone Cannot Do," 205.

8. Krueger, "Effects of Television Viewing," 19, 20.

9. Negroponte, *Being Digital*, 215.

10. Ibid., 227–228.

11. Ibid., 229.

12. Minsky, "Will Robots Inherit the Earth?" 109, 111, 113.

13. Wauzzinski, "Technological Optimism," 149–150.

14. Ibid., 150.

15. Cougar, "Corporations and Overconsumption," 16.

16. Ibid., 17.

17. Ibid.

18. Kates, "Sustaining Life on Earth," 114.

19. Ibid.

CHAPTER 3: POPULATION DENSITY
AND A SUSTAINABLE ENVIRONMENT

1. That God did not address the wives and daughters indicates that the Genesis creation story cannot be taken as the full story, at least not on the topic of human population growth.

2. Gore, *Earth in the Balance*, 31. For an excellent recent discussion of the

population-environment issue from a historical perspective including population-tion projections, see Kates, "Population, Technology, and the Human Environ-ment."

3. The philosophical "rights" of men and women to have offspring have not been extensively evaluated from our perspective. As will be seen, most world religions have assumed that unlimited production of humans is an inherent inalienable and natural right.

4. An eighteenth-century pastor, Thomas Malthus was among the first to warn that the unchecked increase of humankind would some day cause disaster because food production could only increase arithmetically while human populations could expand geometrically.

5. For a review of this discussion, see Cohen, *How Many People?* esp. the "sanctity of the right of reproduction."

6. This holds without even considering the quality of life involved. The strongest case for this argument was made in 1972 by Meadows et al., *The Limits to Growth*, now largely vindicated.

7. Kates, "Population, Technology, and the Environment," 65.

8. This maximum number could well be reached in two generations, at present rates of increase. See ibid. See also Livernash and Rodenburg, "Population, Change, Resources, and Environment."

9. Ehrlich and Ehrlich, quoted in Kates, "Population, Technology, and the Environment," 64. The position advocating negative population growth (NPG) derives from this argument.

10. It is almost gratuitous to remind readers of the daily newspaper reports of encroachment, by suburban sprawl and developers, on the natural habitat, from the rain forests of Brazil, to the forests of western Oregon, to the Chesapeake Bay. The protests of environmental groups seem almost irrelevant and powerless.

11. Bogue, in Micklin, *Population, Environment, and Social Organization*, 344. The ideological rejection of environmental degradation continues, illustrated recently by Esterbrook, *A Moment on the Earth*. He rejects any thought that the environment is under stress: "Is nature really on the run? Several important indexes suggest that it is not." Esterbrook, "Dance of the Ages," 17. Kates cites experts who state that "there is a two to one chance that in the year 2100, global population will fall somewhere between five billion and twenty billion people." Kates, "Population, Technology, and the Environment," 65.

12. Ehrlich, *The Population Bomb;* Meadows et al., *The Limits to Growth*.

13. Micklin, *Population, Environment, and Social Organization*, 123.

14. Corson, "Toward a Sustainable Future."

15. Ibid., 314. In our discussion, *population density* refers to "numbers of peo-

ple in an area relative to its resources and the capacity of the environment to sustain human activities." Ehrlich and Ehrlich, *The Population Explosion*, 38.

16. Data largely from Miller, *Living in the Environment*. See esp. chaps. 11, 13, and 20–24. Additional excellent references include Brown, *State of the World;* Brown, *Vital Signs, 1992;* Piel, *Managing Planet Earth;* and Corson, *The Global Ecology Handbook.*

17. Streams in Pennsylvania are now unsafe because of possible *Giardia* contamination.

18. Although the world spends about $3 billion/day on military programs, only about half this amount is spent toward helping people to have clean drinking water and improved sanitation.

19. Data in this section are largely from Miller, *Living in the Environment*.

20. Note the effects of the atomic bombs dropped on Japan in 1945 and the accident at Chernobyl in 1986, both of which resulted in an 80% increase in thyroid cancer in children and a considerable increase (1.5-fold) in leukemia.

21. "Today humanity is experiencing an epidemic of epidemics," which is due largely to the transmission of diseases through increased global interaction and population density. Platt, "Confronting Infectious Diseases," 114ff.

22. Hardin, "The Tragedy of the Commons." See also Crowe, "The Tragedy of the Commons Revisited."

23. "Unlike the waste from nature's work, the waste from human industry is not 'food' at all. In fact it is poison." McDonough and Braungart, "The Next Industrial Revolution," 88.

24. The "optimistic" position argues that there are two factors in overpopulation: (1) maldistribution (many areas of the world are still unpopulated) and (2) poor planning and organization, which could solve many problems. But the impact of overpopulation on the natural environment itself is ignored.

25. Some scholars and groups have long advocated zero population growth (ZPG), whereas others have advocated negative population growth (NPG).

26. In the Christian religion, e.g., promotion of the production of offspring is derived from Old Testament commands to fill the earth and descriptions of the "blessings" of offspring. The New Testament, however, is remarkably low key about promoting the procreation of children. It is not possible to deal with the variety of positions within Christiandom on the issue of having children, but see Bratton, *Six Billion and More*, for a very creative analysis of the philosophy of "childbirthing" and methods to reduce population growth from an evangelical Christian perspective.

27. However, some religions, such as Buddhism and Christianity, have also promoted celibacy.

28. Aquinas, *The Summa Theologica*, 1:517.

29. This approach, called *functionalism*, proposes that social mores, norms, and practices persist because they must have contributed to the survival of society, else they themselves would not have survived. This position has some weaknesses, thoroughly discussed in the scientific literature, but it is useful when its limits are understood. For a recent discussion see Turner, *The Structure of Sociological Theory*.

30. Davis, *Human Society*, 394–395. This point of view does not assume that the traditional forms of marriage and family are the only forms that could have functioned to produce similar consequences; other forms of institutions could easily have served similar needs.

31. Freedman, "Norms for Family Size," 178. Bratton, *Six Billion and More*, maintains that the subordinate status of women is another basis for many children and that the liberation of women will dramatically reduce rates of childbirth (159–161 passim).

32. This position lies beneath Malthus's thesis that the biological reproduction rate is the maximum possible in relation to the means of subsistence.

33. Goode, *Religion among the Primitives*, 161.

34. Davis, *Human Society*, 561.

35. Davis, quoted in Micklin, *Population, Environment, and Social Organization*, 356. It is mystifying that so few people have questioned this position.

36. Even if this were true, it is now fairly clear that it will come too late to preclude irreparable damage to the planet. See, e.g., McKibbon, "A Special Moment in History," who maintains that if the population explosion is not stabilized by 2020 it will be too late.

37. Hardin, "The Tragedy of the Commons," 1244.

38. Dialogues of Plato: *Republic*, 5:362.

39. China, of course, is a conspicuous exception and for obvious reasons— it is a command control society, and it has a serious overpopulation problem.

40. By *secular value system* is meant the encouragement to satisfy the desire for personal gain and profit free from community and religious restraint.

41. Schnaiberg, *The Environment*, 216.

42. Ewen, *Captains of Consciousness*, 24–25; Henry, *Culture against Man*, 95, 70. Henry proposes that Western society has developed a "pecuniary conception of man," in which the first commandment is "create more desire" and the second is "thou shalt consume" (10–20).

43. Henry, *Culture against Man*, 16.

44. Ehrlich and Holdrin, quoted in Micklin, *Population, Environment, and Social Organization*, 139.

45. Agger, *Gender, Culture, and Power*, 95, 5. Feminism's role in environmentalism is discussed briefly in the conclusion.

46. Bratton, *Six Billion and More*, 171.

47. It is clear why birth control has basically been rejected in free-market societies. A national policy to control births, as in China, is abhorrent.

48. It is intriguing that many religious systems, including Christianity, relegate this issue to the impetuousness or caprice of the deities. Among the Old Colony Mennonites in Mexico, the response to the question, "How many children would you like to have?" was, "As many as God gives us." Redekop, *The Old Colony Mennonites*, 68.

49. The ideas of limits to human production and consumption, as well as human procreation, have been remarkably absent from human thought. The possibility of limits to physical and natural life is a new idea—cf. Solzhenitsyn, *From under the Rubble*, 135–139. Meadows et al., *The Limits to Growth*, very laconically say, "Achieving a self-imposed limitation to growth would require much effort. It would tax the ingenuity, the flexibility, and the self-discipline of the human race" (170).

50. The idea that quantity affects quality and vice versa has only slowly made its way into scientific and philosophical thinking.

51. Henderson, *The Politics of the Solar Age*, 334.

52. There are people who believe human biological life can be extended to 150 years, but this would have major consequences on population density.

53. Henderson, *The Politics of the Solar Age*, 6.

54. O'Conner, *Is Capitalism Sustainable?* 2. The literature on this topic is voluminous and still growing. One early and most perceptive work is Polanyi's *The Great Transformation*. A more recent commentator is Robertson, *The Sane Alternative*. A balanced and extended analysis of the production/consumption syndrome premised on the growth ideology is Schnaiberg, *The Environment*.

55. Postel, "Protecting Forests," 115.

56. Carson, *Silent Spring*. This has been termed the *environmental canary*.

57. Henderson, *The Politics of the Solar Age*, 324.

58. *Development* is one of the most confusing and contradictory words in use in the West and illustrates how ideology can make words contradict their actual meaning. See chapter 1 for an expanded discussion of this idea.

59. A skeptic may respond by thinking, "Even if it is true, why worry about it until the limit is actually reached?" Many ecologists say that the quality of the ecosystem may already have been irreparably damaged: "By this standard, the entire planet and virtually every nation is already vastly overpopulated." Ehrlich and Ehrlich, *The Population Explosion*, 38.

60. Solzhenitsyn, *From under the Rubble*, 135, 137.

61. Ibid., 108, 137–138.

62. It is hardly necessary to point out that societal order has always been

based on the inherent necessity of limits. Ironically, in the Christian tradition there has been a remarkable silence on the question of limits to human reproduction.

63. The technology sector of human culture is probably most deeply committed to the ideology of unlimited progress and advance. Skepticism of technology's "unlimited" nature and possibilities has been met with near derision. Increasingly, however, voices have begun to question this premise and suggest that technology is not amoral. Chapter 2 by Kenton Brubaker suggests that technology's optimism has definite moral influences and implications.

64. The Hutterites have produced some of the highest reproductive rates in the world. In traditions where controlling or limiting births by any means other than abstinence is taboo, adopting a position that supports limiting births will be very difficult. For an excellent treatment of evangelical Protestantism and the overpopulation problem, see Ball, "Evangelicals, Population, and the Ecological Crisis."

65. Brown, "Stopping Population Growth," 202.

66. Corson, "Toward a Sustainable Future," 314.

67. Hauser, *The Population Dilemma*, 183.

CHAPTER 4: GOD'S SPIRIT AND A THEOLOGY FOR LIVING

1. Mellinger, cited in Correll, *Dar schweizerische Taeufermennonitentum*, 125–126.

2. Schreiber, *Our Amish Neighbors*.

3. Ibid., 186–187, 188.

4. Lowdermilk, *Conquest of the Land*, 4.

5. Ibid., 30.

6. Langin, *Plain and Amish*, 56.

CHAPTER 5: MENNONITES, ECONOMICS, AND THE CARE OF CREATION

1. Thigpen, "Savior of the Green Hell."

2. See Redekop and Stahl, "Evangelization of the Native Tribes." Quotation from Thigpen, "Savior of the Green Hell," 14.

3. Yoder, "The 1982 Mennonite Census"; Kauffman and Driedger, *Mennonite Mosaic*.

4. Kollmorgen, *A Contemporary Rural Community*; Fretz, "Farming in North America"; Hostetler, *Amish Society*; Yoder, "Farming in France."

5. Driedger, "Farming in West Prussia and East Prussia." For a review of

Mennonite views on the work ethic and relationships to the land, with relevant bibliography, see Redekop, "Mennonites, Creation, and Work."

6. Redckop, Ainlay, and Siemens, *Mennonite Entrepreneurs.*

7. Nafziger, "Economics."

8. Kennell, interview.

9. Cash prices for corn, wheat, and soybeans in the summer of 1996 were abnormally high, but few farmers or economists expected these to continue. They were a temporary phenomenon due to an abnormally small corn harvest in 1995 and the 1996 drought affecting the 1996 wheat crop in the Southern Plains. By late December 1999 the prices for the three crops listed above had sunk to their lowest level in more than two decades.

10. "Holy Disturbance" by Robert Yoder (1996), used with permission.

11. Yoder, "The Family Farmer."

12. Napier and Sommers, "Farm Production Systems," 74, 75.

13. Kennell, interview.

14. Dyck, "From Airy-Fairy Ideas to Concrete Realities."

15. Dyck, interview.

16. Ibid.

17. Stoltzfus, "Amish Agriculture."

18. Kraybill and Nolt, *Amish Enterprises;* Kanagy and Kraybill, "The Rise of Entrepreneurship."

19. Bender, "Business in North America"; Bender, "Business in Germany"; Yoder, "Business in France and Switzerland"; Krahn, "Business in Russia."

20. Redekop, Ainlay, and Siemens, *Mennonite Entrepreneurs;* Driedger, "Individual Freedom vs. Community Control."

21. Bender, "Business in North America."

22. Sommer, "Emanuel E. Mullet."

23. The Steiner Corp. has been sold, but a family successor has been formed. This is only a representative listing. There is no complete information, but *Mennonite Business and Professional People's Directory, 1978,* produced by MEDA, is a relatively accurate listing of all Mennonite professional and business organizations at the time.

24. Loewen, interview.

25. DeFehr, interview.

26. Loewen, interview.

27. Schlatter, interview.

28. Ibid.

29. Lantz, interview.

30. Redekop, electronic mail letter.

31. Ibid.

32. Stutzman, interview. We remind readers that this survey of Mennonite manufacturing has been cursory and is based on very limited data.

33. Yoder, "Findings from the 1982 Mennonite Census."

34. Janzen Longacre, *The More with Less Cookbook* and *Living More with Less;* Handrich Schlabach, *Extending the Table.*

35. I live in a city with no organized Mennonite congregation, yet we have a self-help store that sells goods distributed through MCC, as well as similar goods distributed through Christian Reformed, Church of the Brethren, and Presbyterian agencies.

36. *Mennonite Reporter,* 24 June 1996, 13.

37. *Mennonite Reporter,* 29 April 1996, 4.

38. Unfortunately, I did not receive much information from Bluffton and Tabor Colleges, but I got quite a bit from Goshen, Messiah, and Fresno Pacific and a fair amount from Bethel and Eastern Mennonite.

39. Leonard, "From Smashed Pallets to Construction Blocks."

40. My teenage daughter is so conscientious that she separates cellophane windows from business envelopes so that they, too, may be recycled. She got most of her motivation from what she learned at a public school with no Mennonite teachers or peers.

41. Hart, electronic mail letter.

42. Miller, interview; Smucker, electronic mail letter; Yoder, electronic mail letter.

43. Ewert, electronic mail letter.

44. Grove, electronic mail letter.

45. Yoder, electronic mail letter.

46. Traub, electronic mail letter.

47. Van Dyke et al., *Redeeming Creation.*

48. Au Sable Institute, Official Bulletin 16.

49. Schrock-Shenk, electronic mail letter.

50. It is revealing that Mennonite Mutual Aid has named its new social responsibility investments Praxis funds!

CHAPTER 6: THE MENNONITE POLITICAL WITNESS
TO THE CARE OF CREATION

1. Androes, interview. The notes for all interviews cited in this chapter are in the private possession of the author.

2. *Lorraine Avenue Messenger* 31 (25 October 1982): 2.

3. See esp. chap. 10, "Make Public Policy Work for Justice," by Roger Claassen, and also chap. 11, "Farmers of the World Connected," by Robert O. Epp, in Platt, *Hope for the Family Farm.*

4. Anecdotal examples will often be used in the absence of exhaustive empirical data.

5. This argument was suggested to me a number of times in the course of my research for this chapter.

6. Information taken from "Resolution on Proposed Land Purchase by the Fort Riley Military Installation," adopted by unanimous voice vote of the conference of delegates at Hesston Mennonite Church, Hesston, Kans., 15 July 1989. See app. 1. See also Meyer, *Christianity and the Environment.*

7. The growth of the retirement home industry is only part of the phenomenal population growth in Lancaster County. During the eighties, Lancaster County grew at a rate of 11.4%, compared to a statewide growth rate of 0.6%. From 1980 to 1987, 58% of the growth of the entire state occurred in Lancaster County alone. Lestz, *Lancaster County's Firsts and Bests*, 4–5, quoted in Testa, *After the Fire*, 142–143. See Testa's book for a personal narrative rendering of Lancaster County issues. According to my informants, Testa's account is essentially accurate, though he has changed the names of Amish people to protect their privacy.

8. Edna's small physical stature belies her large will and determination. Ellis–van Creveld syndrome (EVC), or dwarfism, is extremely common among Lancaster County Amish. See Hostetler, *Amish Society*, 239.

9. The Mill Creek Valley material is from interviews by the author with Lorna Stoltzfus, Doris Goehring, and Edna Esh on 25 March 1996 and a telephone interview with Lorna Stoltzfus on 29 August 1996.

10. "Kickapoo Valley Leads Assault."

11. Boyer, interview.

12. *Wisconsin State Journal*, 8 October 1995.

13. Thomas, interview.

14. Curriculum vitae and notes sent to the author, August 1996.

15. Ortman, "More Ferrets and More Friends."

16. Ortman, letter to the author, 1.

17. Ibid., 2, 5. See also Ortman, "What Good Is a Church?" 1.

18. See Senner, "Making a Difference," 117–119.

19. Senner, interview.

20. Raffensperger, interview.

21. Eight Amish men recently challenged a Wisconsin law requiring them to display a red-and-orange triangle on their horse-drawn buggies. They took

the case to the Wisconsin Supreme Court and won. The Amish have won similar cases in Minnesota, Michigan, and Ohio. "Amish Win Court Ruling."

22. The 1995 *Mennonite Yearbook* lists twenty-eight Anabaptist congregations in Wisconsin, most of them small and many only peripherally related to larger Mennonite church bodies, such as area conferences. Horsch, *Mennonite Yearbook and Directory, 1995*, 42.

23. Boyer, interview.

24. The list of congressional districts by size of Mennonite population was prepared by Karl Shelly of the MCC U.S. Peace Section Office, 100 Maryland Avenue NE, Washington, D.C. 20002. Ratings are found in the *Scorecard*, published by the League of Conservation Voters, 1707 L Street NW, Suite 750, Washington, D.C. 20036.

25. McCarthy, "The Noah Movement." See also "Tending God's Garden."

26. McCarthy, "The Noah Movement." See also app. A, Daryl Byler's letter dated 5 October 1995.

27. See app. B, point 6 under "Commitments of MCC."

28. Bailey, "A Biblical and Theological Apologetic." Bailey was quoting Neff, "Biblical Basis for Political Advocacy," 201.

29. Shelly, "On the Road to Jairus' House," 12.

CHAPTER 7: CREATION, THE FALL, AND HUMANITY'S ROLE IN THE ECOSYSTEM

1. Lynn White's critique of the values toward nature in biblical traditions, which touched off the modern debate about ecology and the Bible, was aimed exclusively at the creation stories in Genesis 1–3. White, "The Historical Roots."

2. A recent discussion of Paul's treatment of this topic is found in Barr, *The Garden of Eden*, 1–20. Barr shows that Paul's understanding of death entering the world through Adam's sin already existed in the Jewish interpretation of his time (e.g., Wisd. of Sol. 2:23).

3. The phrase is Gottwald's in *The Hebrew Bible*, 473.

4. On the relationship between Priestly dietary regulations and the structure of the Priestly creation account, see Douglas, *Purity and Danger*, 41–57, and Eilberg-Schwartz, "Creation and Classification in Judaism."

5. A recent discussion of the "image of God" as a designation of function rather than essence is found in Bird, "Male and Female He Created Them."

6. My interpretation of the Yahwist here is worked out in more detail in *The Yahwist's Landscape*.

7. A recent critique by an ecologist of the notion of the redemption of nature is found in Rolston, "Does Nature Need to Be Redeemed?"

8. This and other examples of corporate irresponsibility are described by Hawken in *The Ecology of Commerce*, 121–136, esp. 129–130.

9. Berry, *The Unsettling of America*, 97, 98, 94.

CHAPTER 8: THE NEW TESTAMENT AND THE ENVIRONMENT

1. Baker, "Biblical Views of Nature," 20.

2. Van Leeuwen, "Christ's Resurrection," 60.

3. Ibid., 58.

4. Zerbe, "The Kingdom of God," 73.

5. As H. Paul Santmire notes, "The God with whom Jesus evidently shared such intimate communion, with whom he identified himself profoundly when he called him *Abba*, was after all, in Jesus' eyes and in the eyes of Jesus' first followers, not only the God of individual souls, not only the God of historical peoples such as the Hebrews, he was also the Maker of Heaven and Earth, the gracious and powerful Creator and Consummator of the whole creation." Santmire, *The Travail of Nature*, 201.

6. Cf. the texts that refer to God as "Lord of heaven and earth" (Matt. 11:25; Luke 10:21; Acts 17:24) and those that speak of heaven as God's "throne" and the earth as God's "footstool" (Matt. 5:35, 23:22; Acts 7:49; Isa. 66:1).

7. Thus, 1 Cor. 8:6; Heb. 1:2; John 1:2–3; Col. 1:15–16; Rev. 3:14.

8. Wilkinson, "Christ as Creator and Redeemer," 29. On the motif of incarnation as portrayed in the Gospel of John, see the carefully argued monograph by Thompson, *The Humanity of Jesus*.

9. Habgood, "Sacramental Approach to Environmental Issues," 48.

10. Consult a concordance for John's use of *semeion/semeia* (sign/signs). For John's accounts of the "signs" themselves, see 2:1–11; 4:46–54; 5:1–9; 6:1–15, 16–25; 9:1–7; 11:1–44; 21:1–14.

11. "We are called as natural beings placed by God in a natural world—and the whole natural process is intended as a vessel of grace. The bread and wine becoming the vessel of promise reveals to us what all of nature, including our own bodies, is designed to be." Hefner, "The Sacramental Paradigm of Nature," 200.

12. Briere, "Creation, Incarnation and Transfiguration," 35.

13. Cf. the comments of Paulos Mar Gregorios: "Christ the Incarnate One assumed flesh—organic, human flesh; he was nurtured by air and water, vegetables and meat, like the rest of us. He took matter into himself, so matter is not alien to him now." Gregorios, "New Testament Foundations," 43.

14. Briere, "Creation, Incarnation and Transfiguration," 37, citing St. John of Damascus, *On the Divine Images*, 23.

15. While Paul is clearly using the language of "planting," "watering," and "giving growth" in a metaphorical sense, it is likewise clear that Paul believes the same to be true in the natural world.

16. Owner of a "field" (1 Cor. 3:9); "landowner who plants a vineyard" (Matt. 21:33; Mark 12:1; Luke 20:9); "vinegrower" who "prunes" fruitful branches and "removes" fruitless ones (John 15:1–2); "landowner" who "hires laborers for his vineyard" (Matt. 20:1); "lord of the harvest" who "sends out laborers into his harvest" (Matt. 9:38; Luke 10:2); grower who "grafts" new branches onto old trees (Rom. 11:17, 19, 23, 24).

17. That John asks the reader to view God's ongoing "work" as the work of "creation" is implicit in the "lifegiving" nature of the healing act that Jesus has just carried out. Cf. the comments of Elsdon in *Bent World*, 134: "We must beware of imagining that God has ceased working in nature; his creative and renewing activity is involved in the world at all levels."

18. As opposed to Santmire, who concludes in *The Travail of Nature* that "John has very little positive regard for human history in general or for the biophysical world as such" (213).

19. The Synoptic Gospels and Acts refer to Jesus' miracles as *dynameis* (acts of power) (Matt. 7:22; 11:20, 21, 23; 13:54, 58; 14:2; Mark 6:2, 5, 14; Luke 10:13; 19:37; Acts 2:22), while the Gospel of John refers to them as *semeia* (signs) (see note 10 above). Note as well Mark's somewhat enigmatic indication that, during the time of Jesus' temptation by Satan in the wilderness, "he was with the wild beasts" (Mark 1:13), an apparent reference to Jesus' ability to coexist safely with otherwise life-threatening creatures of the wild.

20. Paul here speaks metaphorically of Apollos as the one who "waters" that which Paul himself has "planted." See note 15 above.

21. Conversely, when they do not "bear fruit," they are viewed as failing to realize their potential (Matt. 21:18–20) and are subject to the "ax" and "fire" of destruction (Matt. 3:10; Luke 13:9; John 15:6). See also the texts that depict "weeds" (Matt. 13:30) and "chaff" (Matt. 3:12) as destined to be "burned with . . . fire."

22. See, e.g., the references to seeds growing into stalks of wheat (Mark 4:28; John 12:24; cf. 1 Cor. 15:37), vines yielding grapes (John 15:2, 4), and fig trees bearing figs (Luke 13:9). As Jesus summarized it, "every good tree bears good fruit" (Matt. 7:17).

23. Elsewhere James refers to the "bridles" (1:26; 3:2) and "bits" (3:3) that are utilized with horses to "make them obey" humans (3:3).

24. See also the portraits of horses and riders found throughout Revelation (6:2, 4, 5, 8; 9:7, 9, 17; 19:11, 18, 21; cf. 14:20; 18:13).

25. Cf. the comments of John Hart, who identifies the recurrence of the Old Testament motifs of "sabbatical year," "jubilee year," and "gleaning" in the teachings of Jesus. Hart, *The Spirit of the Earth*, 71–81.

26. But note as well several significantly positive references to earthquakes. These "seismic" actions signal the redemptive power of God at work through the death (Matt. 27:54) and the resurrection (Matt. 28:2) of Jesus and the saving power of God at work in releasing imprisoned disciples from their chains and opening their prison doors (Acts 16:26).

27. Cf. the similar birthing language used elsewhere in the Gospels and Paul to describe the cosmic turbulence preceding the Parousia, or the "day of the Lord" (Mark 13:5–8; Matt. 24:5–8; 1 Thess. 5:1–3).

28. Cf. the comments of Carolyn Thomas: "Since there is a solidarity of all created things, Christians cannot be indifferent to the decaying world. Humanity's renewal must be matched by the renewal of all creation (8:19). The implication of Paul's teaching is that Christians are accountable for the rest of creation, for they share solidarity with it in the struggle toward God's final triumph over evil." Thomas, "Romans 8," 34.

29. But note the closely parallel apocalyptic portrayals found within the Synoptic Gospels (Mark 13:1–37; Matt. 24:1–31; Luke 21:5–28). Here there is reference to "war" (Mark 13:7–8), war-driven brutality and devastation (Mark 13:14–20), "famine" (Mark 13:8), "earthquakes" (Mark 13:8), "plagues" (Luke 21:11), "dreadful portents and great signs from heaven" (Luke 21:11), the "darkening" of the sun and moon (Mark 13:24), stars "falling from heaven" (Mark 13:25), the "roaring of the sea and the waves" (Luke 21:25), and the "shaking of the powers of the heavens" (Mark 13:25).

30. Cf. Paul's formulation of this principle in his letter to the Galatians: "Do not be deceived; God is not mocked. For you reap whatever you sow. If you sow to your own flesh, you will reap corruption from the flesh; but if you sow to the Spirit, you will reap eternal life from the Spirit" (Gal. 6:7–8).

31. Cf. Paul's comments to the same effect in his letter to the Romans: "For though they knew God, they did not honor him as God or give thanks to him; but they became futile in their thinking, and their senseless minds were darkened. Claiming to be wise, they became fools; and they exchanged the glory of the immortal God for images resembling a mortal human being or birds or four-footed animals or reptiles" (Rom. 1:21–23).

32. Thus, 2:14, 20, 21; 9:21; 14:8; 17:1, 2, 4, 5, 15, 16; 18:3, 9; 19:2. Cf. the comments of Wesley Granberg-Michaelson: "The heart of the distortion that afflicts all creation is the blasphemous rebellion by humanity, which seeks to be god over the creation. The 'fall' is not a biological fact, but a fracturing of relationships." Granberg-Michaelson, *A Worldly Spirituality*, 112.

33. Thus, 6:10, 11:18, 14:7; 16:5, 7; 18:8, 10, 20; 19:2, 11; 20:12, 13. Cf. the comments of William C. French: "To see God's grace acting in nature also means seeing in ecological destruction a sign of divine judgment and anger." French, "Ecological Degradation," 22.

34. Thus, Rev. 2:5, 16, 21, 22; 3:3, 19; 9:20, 21; 16:9, 11.

35. Granberg-Michaelson, *A Worldly Spirituality*, 111–112.

36. On this point, see Zerbe, "The Kingdom of God," 89–90.

37. Elsewhere Paul moves future re-creation into the present and cosmic re-creation into the personal with his dramatic announcement that, "if anyone is in Christ, there is a new creation: everything old has passed away; see, everything has become new!" (2 Cor. 5:17).

38. On this motif see the helpful discussions of Wilkinson, "Christ as Creator and Redeemer," 39–41, and Manahan, "Christ as the Second Adam," 45–56.

39. Wilkinson, "Christ as Creator and Redeemer," 39. See also the comments of Kehm, "Priest of Creation," 135, and Zerbe, "The Kingdom of God," 88.

40. As Hefner comments, "The resurrection of Jesus that we celebrate is the resurrection of the body." Hefner, "The Sacramental Paradigm of Nature," 198. Van Lecuwen notes that, "if Christ in his death wiped out evil and death, in his resurrection he vindicated the goodness of creation." Van Leeuwen, "Christ's Resurrection," 61.

41. Thus also Zerbe, "The Kingdom of God," 90.

42. Wilkinson, "The Green Agenda," 20.

CHAPTER 9: PACIFISM, NONVIOLENCE, AND THE PEACEFUL REIGN OF GOD

1. The two books by Baker are *Adventures in Contentment* (1909) and *The Friendly Road* (1910). I used the 25th and 27th editions of these works, respectively, printed in 1936.

2. The English word *nonresistance* as used in the Mennonite tradition to identify the rejection of violence against others is an unfortunate translation of the Dutch word *weerlos*. The word actually means "without weapons."

3. Wiebe, *The Temptations of Big Bear*, 29.

4. When I argue that Mennonites took up agriculture as a survival strategy, I am not making a negative judgment. The determination of our ancestors to survive was a perfectly legitimate and honorable response to the dangers they were facing. My concern is to subject to criticism our tendency to read later attitudes into the beginning and so to romanticize our history. I put forward an

explanation, not a censure. Until I am presented with evidence to the contrary, this remains my position.

5. Orland Gingerich suggested, as the reason for the lack of biblical basis for a nonresistant attitude toward the creation among Mennonites, that there is no New Testament word to which they could appeal. Since the New Testament, which they read to the virtual exclusion of the Old Testament, contained no words for the love of the land, they did not articulate it. This is so obvious that I am embarrassed not to have thought of it. It is therefore entirely possible that there was among Mennonites a love of the land but that it was implicit rather than explicit.

This does not, however, remove the problem with which I am concerned. An unarticulated attitude of love and respect for the land could function well, let us say, until the beginning of the twentieth century, by which time the industrialization of agriculture was clearly visible, especially in North America, where most of the descendants of the European Mennonites now live. The lack of biblical and theological articulation of the need to take care of the land because this is part of loving God meant that Mennonites were without defense against the industrialization of agriculture and, as it turned out, followed the rules of industrial agriculture as dictated by the culture rather than a biblical imperative to take care of the land.

6. Ricard, "This Hippie Went to Market," 41.

7. *Mennonite Hymnary*, no. 335.

8. Ibid., no. 332.

9. Tom Finger raised the question about the difference between a stance of *Gelassenheit* and passivism. There is a long tradition of *Gelassenheit* within Christianity. Neither the Rhenish mystics nor the Anabaptists who adopted *Gelassenheit* as a faith stance could be charged with being passive. For those, like myself, who believe *Gelassenheit*, as I describe it in this chapter, to be a legitimate Christian stance in the present, the main temptation is passivism. It can lead to nonaction, even as the chief temptation of committed activists for peace and justice is cynicism and despair. But just because temptations are linked to adopting a stance of *Gelassenheit*, they cannot be a reason to disqualify it as a legitimate Christian stance, even as the temptations to cynicism and despair cannot disqualify activism for peace and justice.

10. Helwig, "One Step from an Old Dance."

CHAPTER 10: AN ANABAPTIST/MENNONITE THEOLOGY OF CREATION

1. Robert Friedmann's *Theology of Anabaptism* includes no discussion of creation.

2. See Loewen's collection of Mennonite confessions in *One Lord, One Church*. Creation is so marginal that it does not appear in Loewen's list of the twenty-one most frequently appearing articles (p. 267, but cf. 260–265, 268). A notable exception is the Ris Confession (1766), which infers characteristics of God from the created order (87).

3. E.g., the Schleitheim Confession of 1527, article 4: "There is nothing else in the world and all creation than good or evil, believing and unbelieving, darkness and light, the world and those who are [come] out of the world, God's temple and idols, Christ and Belial, and none will have part with the other." Loewen, *One Lord, One Church*, 80.

4. Dirk Philips, e.g., interpreted the "new heavens" as realities already existing spiritually as "believers in whom God dwells" and the "new earth" to be presently "the hearts of the Christians." Finally, however, all believers will inherit the new heavens and earth when "the perfect transformation out of the perishable into that eternal imperishable and glorified takes place." Dyck, Keeney, and Beachy, *The Writings of Dirk Philips*, 318–319. Much research remains to be done, however, to determine whether Philips and others really expected the final state to be entirely spiritual or the present creation to be wholly destroyed.

5. Redekop, "Toward a Mennonite Theology."

6. Friedmann placed an "ontological dualism" at the heart of Anabaptism but described it as follows: "an uncompromising ontological dualism in which Christian values are held in sharp contrast to the values of the 'world' in its corrupt state." Friedmann, *The Theology of Anabaptism*, 38. It is unfortunate that Friedmann used the term *ontological*, which is appropriately applied to dualisms like those between matter and spirit, when his own words show that he probably meant an ethical and religious dualism.

7. Redekop, "Toward a Mennonite Theology," 403.

8. Redekop, "Mennonites, Creation, and Work," 348–366, 353–357. Redekop maintains that most Mennonites have now entered a third phase, dominated by the "entrepreneurial motif," where they view land chiefly as a source of production and profit (357–361).

9. See Redekop, "Toward a Mennonite Theology," 397–400. Walter Klaassen suggests that the virtue of *Gelassenheit* (i.e., totally submitting to God's will) can guide an environmental ethic. However, the major mystics in the tradition from which Anabaptism drew this notion (Meister Eckhart, Johann Tauler, Henry Suso, etc.) seem to me to be more negative about the created order than Klaassen claims.

10. See, e.g., Anderson, *Creation versus Chaos*, 11–77; Young, *Creator, Creation, and Faith*, 25–63; Reumann, *Creation and New Creation*, 31–42.

11. Norman Kraus, in *God Our Savior*, devotes his first chapter to Jesus and

his saving work before speaking of God in general in chap. 3 and creation in chap. 4 (107–113).

12. These were proposed by Harold Bender in *The Anabaptist Vision*, but see note 17 below. Although J. Denny Weaver utilizes a more complex approach to Anabaptist origins, he arrives at very similar results. See *Becoming Anabaptist*.

13. In Matt. 10:29, Jesus says that two sparrows are sold for an *assarion* (1/16 of a *denarius*). But in Luke 12:6, he says that five are sold for two *assaria*. Apparently, if one bought "in bulk," a fifth sparrow would be thrown in free. In Luke, Jesus is saying that even the free one "is not forgotten in God's sight."

14. See McFague, *Models of God* and *The Body of God*, and McDaniel, *Earth, Sky, Gods, and Mortals*.

15. See Ruether, *Gaia and God*, and Birch and Cobb, *The Liberation of Life*. This positive interpretation of evolution has been popularized, in a way most environmental theologians accept, by Brian Swimme and Thomas Berry. A corresponding environmental spirituality has been advocated by Matthew Fox, esp. in *Original Blessing* and *The Coming of the Cosmic Christ*. For further discussion of these issues, see Finger, "Trinity, Ecology, and Panentheism" and *Self, Earth, and Society*.

16. Recent attempts to distinguish a historical Jesus from cosmic claims about him and his resurrection have been collectively called the "third quest for the historical Jesus." See, e.g., Funk and Hoover, *The Five Gospels*; Crossan, *The Historical Jesus*; and Borg, *Jesus, Meeting Jesus*, and *Jesus in Contemporary Scholarship*. This movement routinely dismisses cosmic claims about Jesus and his resurrection as products of the early church. By so doing, however, this quest, which aims to be historical, fails to address a fundamental historical question: How could Jesus have been considered in this way by the church if his history showed no basis for it? For a response, see N. T. Wright, *The New Testament*, and Finger, *Self, Earth, and Society*.

17. Beachy, *The Concept of Grace*, and Finger, "Anabaptism and Eastern Orthodoxy," 76–83. Despite the tendency of some to reduce Bender's "Anabaptist Vision" to ethics, it is evident that Bender himself believed in Christ as "the living present Savior who accomplishes our present salvation and continues to be in us and to be in His church." Bender, "Who Is the Lord?" 156; cf. *These Are My People*, 25.

18. Despite understandable reservations about *Father* and *Son* as masculine terms, I retain them because the New Testament employed them to emphasize not gender, but intimacy and fidelity in a relationship. Rather than reject these terms because later Christianity patriarchalized them, I retain them to critique these very perversions from a biblical standpoint. For fuller discussion, see Finger, *Christian Theology*, 2:485–490.

19. See Moltmann, *The Crucified God*, 235–249; Moltmann, *The Trinity and the Kingdom*, 61–96; and my exposition of Moltmann in *Following Jesus Christ*, 7–22.

20. See Moltmann, *The Way of Jesus Christ*, 151–159.

21. 2 Cor. 4:6, 5:17; Gal. 6:15; Eph. 4:24; James 1:18. These texts, however, only begin to express this theme. According to Paul Minear, "the new creation" is one of three main images for God's people, the church, in the New Testament. Minear, *Images of the Church*, 105–135.

22. Affirmations regarded as part of the earliest preaching (kerygma) present Jesus' work as the fulfillment of God's historical plan (Acts 2:23, cf. 13:27; Rom. 1:2; 1 Cor. 15:3–4). Later texts call it the mystery hidden throughout previous generations (Rom. 16:25–26; Eph. 1:9–10, 3:4–6; Col. 1:26–27, 2:2). Others speak of it as destined from before the beginning of the world (1 Peter 1:20, 2 Tim. 1:9–10, Titus 1:2–3, cf. Heb. 9:26).

23. Cf. Kraus, *God Our Savior*, 46.

24. For a fuller discussion, see Finger, *Christian Theology*, 2:77–79, 2:491–492.

25. Botterweck and Ringgren, *Theological Dictionary of the Old Testament*, 2:242–249. Yet Genesis may not be entirely clear on this matter, since 1:2 could indicate that some inchoate matter already existed.

26. Rom. 4:17. Here Paul is describing Abraham's faith, yet his ultimate reference point for describing the God in whom Abraham believed is Jesus' resurrection (4:23–25).

27. Remember, e.g., that their relationships could change in different phases of the redemptive work. For instance, while the Spirit raised the Son, the Son later sent the Spirit. When we view these relationships beginning with creation and reaching to the present (in a *protological* order), it often appears that activity originates with the Father and passes through Son and then Spirit. When we begin with postresurrection experience, however, the Spirit first comes to our attention. In this *eschatological* order, the Spirit points toward the Son, who points toward the Father, in whom all things will finally be summed up (1 Cor. 15:24–28). Traditional theology almost always discusses the Trinity from a protological vantage point, which can easily give the impression of hierarchical intratrinitarian structure. See Moltmann, *The Future of Creation*, 80–96.

28. For a slower discussion, see Finger, *Christian Theology*, 2:379–405, 2:433–455.

29. See Moltmann, *The Trinity and the Kingdom*, 108–111.

30. Classic presentations of this hypothesis are Lovelock, *Gaia*, and Margulis and Sagan, *Microcosmos*.

31. Fox, *A Spirituality Named Compassion*, 33.

32. Ruether, *Gaia and God*, 249, cf. 31, 43, 222; McFague, *Models of God*, 76, 120; McFague, *The Body of God*, 60, 70; McDaniel, *Earth, Sky, Gods, and Mortals*, 166; Birch and Cobb, *The Liberation of Life*, 192.

33. See the discussions in Nitecki, *Evolutionary Progress*. The theme of advance is also severely challenged by the "punctuated equilibria" evolutionary school. See Eldredge, *Time Frames*.

34. All creatures, however, also have an instrumental value and occasionally can and must be used as means to other creatures' ends. Environmental decisions involve the complex interplay between intrinsic and instrumental values. My point here is that instrumental values will not be rightly considered unless some intrinsic value is also affirmed (cf. McDaniel, *Earth, Sky, Gods, and Mortals*, 65–68; Birch and Cobb, *The Liberation of Life*, 141–165).

35. McFague ultimately cannot build her ethic on evolution, but instead builds it on the "destabilizing, inclusive, nonhierarchical vision of fulfillment for all of creation . . . illuminated by the paradigmatic story of Jesus." McFague, *Models of God*, 49. For "what consonance can there possibly be between Christianity's inclusion of the outcasts of society . . . and biological evolution, in which millions are wasted, individuals are sacrificed for the species, and even whole species are wiped out in the blinking of an eye?" McFague, *The Body of God*, 171.

36. I assume that God's creation also includes spiritual realities, such as angels. Throughout this chapter, however, I am focusing on its physical dimensions, and it is these that I signify by *creation*.

37. See Finger, *Christian Theology*, 1:156–158.

38. Some texts that have been interpreted as envisioning complete release of humans from their bodies actually speak of them instead as being clothed (*enduo*, 1 Cor. 15:53–54), or clothed upon (*ependuomai*, 2 Cor. 5:4), by a higher reality.

CHAPTER 11: THE EARTH IS A SONG MADE VISIBLE

1. Hirschfelder and Molin, *Encyclopedia of Native American Religions*, ix, x.

2. O'Brien, *American Indian Tribal Governments*, 37–38.

3. Grebel, "Letter to Müntzer, 1524."

4. Weatherford, *Indian Givers*, 71–72.

5. Giglio, *Southern Cheyenne Women's Songs*, 6.

6. Petter, *Cheyenne Spiritual Songs*, hymn 95.

7. Ibid., hymn 39.

8. *Hymnal: A Worship Book*, hymn 78.

9. Hoig, *Peace Chiefs of the Cheyennes*, 66.

10. Berger, *A Long and Terrible Shadow*, 160–162.

CHAPTER 12: TOWARD AN ANABAPTIST/MENNONITE
ENVIRONMENTAL ETHIC

1. McClendon, *Systematic Theology*, 79.

2. "North Americans make up about 5 percent of the world's population, yet consume slightly more than 27 percent of the world's total commercial energy." Gex, "Has Awareness Brought Lifestyle Changes?" 4.

3. Penner, "Conference Aims to Create Green Theology," 13. Further, according to Klaassen, "Mennonites had not gone into agriculture out of a love for the land, but merely to survive." Ibid. On this issue, see also Redekop, "Mennonites, Creation, and Work."

4. Redekop, "Toward a Mennonite Theology," 395.

5. Ibid., 389. See also Finger, "Kingdom of God," 5:491.

6. Redekop, "Toward a Mennonite Theology," 396. The "world" has been a very difficult concept for Anabaptism. Harold S. Bender states that "a major difficulty has been that of identifying precisely what 'world' is, and therefore what 'worldliness' is." *Mennonite Encyclopedia*, 4:981.

7. Ruth, "America's Anabaptists," 26.

8. See Bender, *The Anabaptist Vision*, and Weaver, *Becoming Anabaptist*.

9. Finger, *Christian Theology*, vol. 2.

10. According to Dennis Miller, "Mennonites were not ready, for the most part, to articulate a theology of the sacramental nature of created things, whereby the Incarnation serves as a testimony that God is redeeming the sin-corrupted creation." Miller, "Theology of Creation," 5:211.

11. Simons, "The Spiritual Resurrection" (1536), 58–59.

12. Redekop, "Toward a Mennonite Theology," 399; Finger, "Kingdom of God"; Redekop, "Toward a Mennonite Theology," 388; Ruth, "America's Anabaptists," 29.

13. McClendon, *Systematic Theology*, 88–89.

14. Ibid., 88.

15. Ibid., 91; Redekop, "Toward a Mennonite Theology," 389.

16. Ruth, "America's Anabaptists," 29. Efforts to stimulate Anabaptist environmentalism from within have been less successful than hoped. The Mennonite Environmental Task Force lost its funding in 1993.

17. Meyer, *Christianity and the Environment*, 3.

18. Ibid.; Good, "The Amish," 28.

19. Kropf, "Let All Things," 11; Rich, "Hope for the Planet," 271; Penner, "Conference Aims," 14.

20. Reiheld, "Living by Soil," 19.

21. Miller, "America's Anabaptists," 31; Gex, "Has Awareness Brought Lifestyle Changes?" 4.

22. Penner, "Conference Aims," 14; Price, "What Vision?" 10; Ruth, "America's Anabaptists," 29.

23. Bicksler, "The Environment," 2–3.

24. Miller, *Our People*, 6–8.

25. Coats, "Tribe Conscious of Environmental Crisis," 3.

26. For one example of Mennonite service efforts that have resulted in environmental degradation, see Redekop and Stahl, "The Impact on the Environment."

27. Schrock-Shenk, "Disasters Are Not Always Natural," 277.

28. Comstock, "Must Mennonites Be Vegetarians?" 273.

29. These thinkers include Art and Jocele Meyer, Tom Finger, Delores Histand Friesen, Milo Kauffman, Walter Klaassen, Calvin Redekop, and Dorothy Jean Weaver. See also the 1991 Mennonite Central Committee publication *Caretakers: Earth Stewardship for Children* and the "Earth Stewardship" packet of educational materials.

30. Kauffman, *Stewards of God*.

31. Meyer and Meyer, *Earthkeepers*; Meyer, *Christianity*, 1.

32. Redekop, "Toward a Mennonite Theology," 395, 394, 399.

33. Redekop, "Toward a Menonite Theology," 393.

34. Comstock, "Must Mennonites Be Vegetarians?"

35. Meyer, *Christianity*, 6; Redekop, "Toward a Mennonite Theology," 403, 400.

36. Comstock, "Must Mennonites Be Vegetarians?" 273; Redekop, "Toward a Mennonite Theology," 401.

37. Weaver, cited in Penner, "Conference Aims," 13–14; Meyer, *Christianity*, 6.

38. Comstock, "Must Mennonites Be Vegetarians?" 273.

39. Klaassen, cited in Penner, "Conference Aims," 13; Redekop, "Toward a Mennonite Theology," 394.

40. Meyer and Meyer, *Earthkeepers*, 58; Redekop, "Toward a Mennonite Theology," 386, 403, 394.

41. Meyer, *Christianity*, 2; Brown, "Just the Beginning," 32.

42. Ruth, "America's Anabaptists," 26, 25.

43. Friesen, *Living More with Less Guide*, 26.

44. Ibid., 99.

45. Koontz, cited in Reimer, "How Does Peace Theology Relate?" 10; Ruth, "America's Anabaptists," 29.

46. Simons, "Blasphemy of John of Leiden" (1535), 43.

47. Meyer, *Christianity*, 7–8.

48. Redekop, "Toward a Mennonite Theology," 403.

49. Ibid., 397; see also Redekop's early "Why a True-Blue Nonresistant Christian"; Comstock, "Must Mennonites Be Vegetarians?" 273.

50. Redekop, "Toward a Mennonite Theology," 392.

51. Miller, "America's Anabaptists," 30.

52. Kreider, cited in Price, "What Vision Will Preserve?" 10.

53. Comstock, "Must Mennonites Be Vegetarians?" 273.

54. Good, "The Amish," 28.

55. Williard, "Amish Farm by Rules," 24; Reiheld, "Living by Soil," 20; Miller, *Our People*, 19.

56. Miller, "America's Anabaptists," 30.

57. Simons, "The Spiritual Resurrection" (1536), 60; Mack, cited in Brown, "Just the Beginning," 32.

58. Miller, *Our People*, 46.

59. Ibid., 6.

60. Williard, "Amish Farm by Rules," 25; Good, "The Amish," 28.

61. Good, "The Amish," 28.

62. After ten or fifteen years, "the wheels may need to be replaced or the interior recovered," but "the wooden framework, the axles, and the steel springs endure" and "are nearly indestructible." Reiheld, "Buggy Travel," 11.

63. Williard, "Amish Farm by Rules," 24–25.

64. Comstock, "Must Mennonites Be Vegetarians?" 273; Simons, "Brief Confession on the Incarnation" (1544), 451.

65. Niebuhr, cited in Ruth, "America's Anabaptists," 29; Ruth, "America's Anabaptists."

66. Miller, *Our People*, 17.

67. Ruth, "America's Anabaptists," 28; see also Miller, *Our People*, 18–19.

68. Kline, cited in Penner, "Conference Aims," 14; Kaufman, cited in Reiheld, "Living by Soil," 20.

69. Gex, "Has Awareness Brought Lifestyle Changes?" 4.

70. Schrock-Shenk, "Disasters Are Not Always Natural," 277.

71. McClendon, *Systematic Theology*, 91.

72. Ibid., 88, 89–90.

73. Gex, "Has Awareness Brought Lifestyle Changes?" 4.

74. Kropf, "Let All Things," 11; McClendon, *Systematic Theology*, 90.

75. Miller, *Our People*, 45–46.

76. Simons, "Epistle to Martin Micron" (1556), 917; Simons, "Reply to Gellius Faber" (1554), 740.

77. Rich, "Hope for the Planet," 270.

78. Ibid., 271.

79. Reimer, "How Does Peace Theology Relate?" 10; Penner, "Conference Aims," 14; Ruth, "America's Anabaptists," 26.

80. Miller, "America's Anabaptists," 30, 32, 33.

81. Miller, "Growing Up Anabaptist," 27; see also Good, "The Amish," 28.

82. Ruth, "America's Anabaptists," 25.

83. Brown, "Just the Beginning," 32.

84. Miller, "America's Anabaptists," 30.

85. Ibid., 31.

86. Simons, "Instructions on Excommunication" (1558), 979, 977.

87. Simons, "The Spiritual Resurrection" (1536), 60.

CHAPTER 13: THE ENVIRONMENTAL CHALLENGE BEFORE US

1. Dubos, "Environment," 2:120. The philosophical question of whether humans can, in fact, separate themselves from the environmental forces that have created them and change the environment in a positive way is being increasingly doubted. "Man cannot perceive the external world objectively without concepts because his perceptual apparatus is shaped by the environment" (127). The authors of this book take the position that a religiously received mandate, because of its basic limitation of human authority, is the only way humans can arrive at a moral and practical understanding of their role in relating to the environment.

2. Ibid., 121.

3. Ibid. "The literature concerned with the [reciprocal] relation of humans and culture to environment will appear in ever-increasing volume, concerned with the meaning and value not only of human but of all life, with the environments that support them" (134).

4. White, "The Historical Roots," 1206; Kearns, "Saving the Creation," 55.

5. This position was especially true of the more evangelical Christian tradition. The Au Sable Institute, headed by Calvin Dewitt, reflects this position— "The problem is not with Christianity, but with not being true to Christianity." Kearns, "Saving the Creation," 59.

6. Ibid., 58. For a classification of the positions evangelical Christianity has taken, see Ball, "Evangelicals, Population, and the Ecological Crisis."

7. Kearns, "Saving the Creation," 59, 60.

8. Ibid., 57.

9. Ball, "Evangelicals, Population, and the Ecological Crisis," 231, 237.

10. Ibid., 60, 61.

11. Adams, *Ecofeminism and the Sacred*, 1, 2.

12. Ibid., 122.

13. This book represents the most focused analysis and synthesis of the Anabaptist environmental ethic until the present; however awareness of the necessity of producing a stance on environmental issues is increasing. It is noteworthy that most Mennonite colleges have just launched environmental majors and undoubtedly the rest will follow.

14. Zerbe, "The Kingdom of God," 74.

15. Ibid., 75–76, 79.

16. White, "The Historical Roots," 155.

17. The warning that the world is becoming overcrowded is an increasing theme recited almost daily in the press and mass media. "People on the street" are becoming aware that we live in a finite world, a negative illustration being the NIMBY (not in my back yard) movement. But the resistance to this reality by the reigning market capitalist growth community will be very great until the evidence becomes overwhelming.

18. But environmental "cleanup" is not always a good thing, for it does not address the behavior of persons who pollute and may even assuage their guilt for polluting, since "someone will always clean it up."

19. McDonough and Braungart, "The Next Industrial Revolution," 88. They maintain that a sustainable world can be achieved only when waste or by-products do not contribute to pollution of the environment but rather are the food for the next generation. A tall order, but they present convincing evidence that such a state can be achieved.

20. Sherrard, "Sacred Cosmology," 8.

21. Dubos continues, "Wherever conditions are suitable for technological development, the earth is losing not only its ecological balance and pristine beauty, but also its fitness for biological and mental health" (122). Readers are encouraged to read the entire article entitled "Environment," for it is one of the best and most succinct statements on the environmental crisis.

22. "The Meditation of Marcus Aurelius," 266.

23. Hut (1527), in Klaassen, *Anabaptism in Outline*, 49ff.

24. Increasingly, thoughtful commentators are saying that the destructive consequences of human behavior regarding the environment will eventually force humans to change their ways, so why wait? A personal embarrassment is the fact that, when my Mennonite ancestors homesteaded in eastern Montana in the early 1900s, they contributed to the devastation of the extremely delicate grass root structure on the prairies, which resulted in the terrible dust bowl of the early 1930s.

Bibliography

Adams, Carol J., ed. *Ecofeminism and the Sacred*. New York: Continuum, 1993.

Agger, Ben. *Gender, Culture, and Power: Toward a Feminist Postmodern Critical Theory*. Westport, Conn.: Prager, 1993.

"Amish Win Court Ruling." *Washington Post*, 6 June 1996.

Anderson, Bernhard. *Creation versus Chaos*. New York: Association Press, 1967.

Androes, Jo. Interview by the author. Telephone, 14 February 1996.

Aquinas, Thomas. *The Summa Theologica*, vol. 1. Chicago: Encyclopedia Britannica, 1952.

Au Sable Institute. Official Bulletin No. 16. Mancelona, Mich.: The Institute, 1996.

Aurelius, Marcus. *The Meditations of Marcus Aurelius*. Chicago: Encylopedia Britannica, 1952.

Bailey, Wilma Ann. "A Biblical and Theological Apologetic for Anabaptist Advocacy." Paper presented at the seminar "Engaging the Powers," Washington, D.C., 2 April 1996.

Baker, John Austin. "Biblical Views of Nature." In *Liberating Life: Contemporary Approaches to Ecological Theology*, edited by Charles Birch, William Eakin, and Jay B. McDaniel, 9–26. Maryknoll, N.Y.: Orbis Books, 1990.

Baker, Ray Stannard. *Adventures in Contentment*, 25th ed. London: Andrew Melrose, 1909.

————. *The Friendly Road*, 27th ed. London: Andrew Melrose, 1910.

Ball, Jim. "Evangelicals, Population, and the Ecological Crisis." *Christian Scholar's Review* 28 (winter 1998): 226–253.

Barr, James. *The Garden of Eden and the Hope of Immortality*. London: SCM, 1992.

Beachy, Alvin. *The Concept of Grace in the Radical Reformation*. Nieuwkoop, The Netherlands: B. DeGraaf, 1977.

Bean, Heather Ann Ackley. *Healthy Home Cooking: Meatless, Dairyfree Options for Everyday Living*. San Dimas, Calif.: Woody Inventions, 1996.

Beasley-Murray, G. R. *The Book of Revelation*. New Century Bible. Greenwood, S.C.: Attic Press, 1974.

Bender, Harold S. *The Anabaptist Vision*. Scottdale, Pa.: Herald Press, 1944.

———. "Business among the Mennonites in Germany." *Mennonite Encyclopedia* 1 (1955): 482–483.

———. "Business among the Mennonites in North America." *Mennonite Encyclopedia* 1 (1955): 483–484.

———. "Farming." *Mennonite Encyclopedia* 2 (1956): 303–313.

———. *These Are My People*. Scottdale, Pa.: Herald Press, 1962.

———. "Who Is the Lord?" *Mennonite Quarterly Review* 38 (April 1964): 152–160.

Bender, Harold S., Winfield Fretz, and John Howard Yoder. "Business." *Mennonite Encyclopedia* 1 (1955): 480–484.

Berger, Thomas R. *A Long and Terrible Shadow*. Vancouver: Douglas & McIntyre, 1991.

Berry, Wendell. *The Unsettling of America*. San Francisco: Sierra Club, 1977.

Bicksler, Harriet Sider. "The Environment: Not Only a White Middle-Class Issue." *MCC Contact* 15 (August 1991): 4–5.

Birch, Charles, and John B. Cobb Jr. *The Liberation of Life: From the Cell to the Community*. Cambridge: Cambridge University Press, 1981.

Birch, Charles, William Eakin, and Jay B. McDaniel, eds. *Liberating Life: Contemporary Approaches to Ecological Theology*. Maryknoll, N.Y.: Orbis Books, 1990.

Bird, Phyllis. "'Male and Female He Created Them': Gen. 1:27b in the Context of the Priestly Account of Creation." *Harvard Theological Review* 74 (1981): 129–159.

Booth, William. "Ecosystem Paradoxically Glows at Former Atomic Bomb Factory Site." *Washington Post*, 26 May 1996.

Borg, Marcus. *Jesus: A New Vision*. San Fransicso: Harper, 1987.

———. *Jesus in Contemporary Scholarship*. Valley Forge, Pa.: Trinity, 1994.

———. *Meeting Jesus Again for the First Time*. San Francisco: Harper, 1994.

Botterweck, Johannes, and Helmer Ringgren. *Theological Dictionary of the Old Testament*, 2:242–249. Grand Rapids, Mich.: Eerdmans, 1975.

Boyer, Dennis. Interview by the author. Telephone, 29 August 1996.

Bratton, Susan Power. *Six Billion and More: Human Population Regulation and Christian Ethics*. Louisville: Westminster/John Knox Press, 1992.

Briere, Elizabeth. "Creation, Incarnation and Transfiguration: Material Creation and Our Understanding of It." *Sobornost* 11, no. 1–2 (1989): 31–40.

Bromfield, Louis. *Pleasant Valley*. Wooster, Ohio: Wooster Book Co., 1945.

Brown, Dale W. "Just the Beginning." *Christianity Today* 34 (22 October 1990): 32.

Brown, Lester. "Stopping Population Growth." In *The State of the World*, edited by Lester Brown, 200–221. New York: W. W. Norton, 1985.

Brown, Lester, ed. *State of the World*. New York: W. W. Norton, 1996.

———, ed. *Vital Signs, 1992*. New York: W. W. Norton, 1992.

Brown, Lester R., Christopher Flavin, and Sandra Postel. *Saving the Planet: How to Shape an Environmentally Sustainable Global Economy*. New York: W. W. Norton, 1991.

Brown, William M. "The Outlook for Future Petroleum Supplies." In *The Resource Earth: A Response to Global 2000*, edited by Julian L. Simon and Herman Kahn, 361–386. Oxford: Basil Blackwell Publishers, 1984.

Bruteau, Beatrice. "Eucharistic Ecology and Ecological Spirituality." *Cross Currents* 40 (winter 1990–91): 499–514.

Burkholder, J. Richard, and Calvin Redekop. *Kingdom, Cross, and Community*. Scottdale, Pa.: Herald Press, 1975.

Burr, Linda. *Mennonites and the Ecological Crisis*. Elkhart, Ind.: AMBS, 1987.

Cairncross, Frances. *Green Inc: A Guide to Business and the Environment*. Washington, D.C.: Island Press, 1995.

Carson, Rachel. *Silent Spring*. New York: Houghton Mifflin, 1962.

Cather, Willa. *The Emerging Voice*. New York: Oxford University Press, 1987.

Charlesworth, Roberta A. *The Second Century Anthology of Verse*. Toronto: Oxford University Press, 1969.

Claassen, Roger. "Make Public Policy Work for Justice." In *Hope for the Family Farm: Trust God and Care for the Family Farm*, edited by LaVonne Godwin Platt, 142–158. Newton, Kans.: Faith & Life Press, 1987.

Coats, Scott. "Tribe Conscious of Environmental Crisis." *MCC Contact* 15 (August 1991): 3–4.

Cobb, Clifford, Ted Halstead, and Jonathan Rowe. "If the GDP Is UP, Why Is America Down?" *Atlantic Monthly*, October 1995, 59–78.

Cohen, Joel E. *How Many People Can the Earth Support?* New York: W. W. Norton, 1995.

Comstock, Gary. "Must Mennonites Be Vegetarians?" *Mennonite* 107 (23 June 1992): 273.

"Conservation/Ecology: A Mennonite Mandate." Video, Mennonite Central Committee, Akron, Pa. (1983).

Correll, Ernst H. *Das schweizerische Taeufermennonitentum*, 125–126. Tuebingen: J. C. B. Mohr (Paul Siebek), 1925.

Corson, Walter H. "Toward a Sustainable Future: Priorities and Progress." In *The Global Ecology Handbook*, chap. 16. Boston: Beacon Press, 1990.

Cougar, Frank. "Corporations and Their Role in Overconsumption." *Green Cross* 2 (summer 1996): 16–17.

Crossan, John Dominic. *The Historical Jesus: The Life of a Mediterranean Jewish Peasant.* San Francisco: Harper, 1991.

Crowe, Beryl L. "The Tragedy of the Commons Revisited." *Science* 166 (28 November 1969): 1103–1107.

Daly, Herman E. *Beyond Growth: The Economics of Sustainable Development.* Boston: Beacon Press, 1996.

Daly, Herman E., and John B. Cobb Jr. *For the Common Good: Redirecting the Economy toward Community, the Environment, and a Sustainable Future.* Boston: Beacon Press, 1989.

Daly, Herman E., and Kenneth E. Townsend, eds. *Valuing the Earth: Economics, Ecology, Ethics.* Cambridge: MIT Press, 1993.

Davis, Kingsley. *Human Society.* New York: Macmillan, 1949.

Deevey, Edward S., Jr. "The Human Population." *Scientific American* 203 (September 1960): 194–204.

DeFehr, Frank. Interview by the author. Telephone, Winnipeg, Manitoba, 26 July 1996.

DeWitt, Calvin B. *Earthwise: A Biblical Response to Environmental Issues.* Grand Rapids, Mich.: CRC Publications, 1994.

———. *The Environment and the Christian: What Does the New Testament Say about the Environment.* Grand Rapids: Baker Book House, 1991.

The Dialogues of Plato. Translated by Benjamin Jowett. Chicago: Encyclopedia Britannica, 1952.

Diamond, Jared. "Easter's End." *Discover* (August, 1995): 63–69.

Disch, Robert, ed. *The Ecological Conscience.* Englewood Cliffs, N.J.: Prentice-Hall, 1970.

Dorfman, Robert, and Nancy S. Dorman, eds. *Economics of the Environment: Selected Readings.* New York: W. W. Norton, 1993.

Douglas, Mary. *Purity and Danger.* London: Routledge & Kegan Paul, 1966.

Driedger, Johannes. "Farming among the Mennonites in West Prussia and East Prussia." *Mennonite Encyclopedia* 2 (1956): 311–313.

Driedger, Leo. "Individual Freedom vs. Community Control." *Journal for the Scientific Study of Religion* 21, no. 3 (1982): 226–241.

———. "Industrialization." *Mennonite Encyclopedia* 5 (1990): 438–439.

———. "Mennonite Business in Winnipeg." In *Anabaptist / Mennonite Faith and Economics,* edited by Calvin Redekop, Vic Krahn, and Samuel J. Steiner, 177–196. Lanham, Md.: University Press of America, 1994.

Dubos, René. "Environment." In *Dictionary of the History of Ideas,* edited by Philip P. Wiener, 2:120–127. New York: Charles Scribner's Sons, 1973.

Durning, Alan. *How Much is Enough? the Consumer Society and the Future.* New York: W. W. Norton, 1992.

Dyck, Bruno. Electronic mail letter. Winnipeg, Manitoba, 20 August 1996.

———. "From Airy-Fairy Ideas to Concrete Realities: The Case of Shared Farming." *Leadership Quarterly* 5, no. 3/4 (1994): 227–246.

Dyck, Cornelius, William Keeney, and Alvin Beachy, trans. and eds. *The Writings of Dirk Philips.* Scottdale, Pa.: Herald Press, 1992.

Ehrlich, Paul. *The Population Bomb.* New York: Ballantine, 1968.

Ehrlich, Paul, and Anne Ehrlich. *The Population Explosion.* New York: Simon & Schuster, 1990.

Ehrlich, Paul, and John P. Holdren. "Population and Panaceas.: A Technological Perspective." In *Population, Environment, and Social Organization: Current Issues in Human Ecology,* edited by Micheal Micklin, 125–143. Hinsdale, Ill.: Dryden Press, 1973.

Eilberg-Schwartz, Howard. "Creation and Classification in Judaism: From Priestly to Rabbinic Conceptions." *History of Religions* 26 (1986–87): 357–381.

Eldredge, Niles. *Time Frames.* New York: Simon & Schuster, 1985.

Elsdon, Ron. *Bent World: A Christian Response to the Environmental Crisis.* Downers Grove, Ill.: InterVarsity Press, 1981.

Epp, Robert. "Farmers of the World Connected." In *Hope for the Family Farm: Trust God and Care for the Family Farm,* edited by LaVonne Godwin Platt, 162–168. Newton, Kans.: Faith & Life Press, 1987.

Epp-Tiessen, Esther. *Altona: The Story of a Prairie Town.* Altona, Manitoba: D. W. Friesen, 1982.

Ernsthafte Christenpflicht. Wooster, Ohio: Johann Sala, 1826.

Esterbrook, Gregg. "Dance of the Ages." *Washington Post Magazine,* 9 April 1995, 17.

———. *A Moment on the Earth.* New York: Viking, 1995.

Ewen, Stuart. *Captains of Consciousness.* New York: McGraw-Hill, 1976.

Ewert, Norman. Electronic mail letter. Wheaton, Ill., 11 July 1996.

"The Final Issue." *Inter-Mennonite Farm Newsletter,* n.d.

Finger, Tom. "Anabaptism and Eastern Orthodoxy: Some Unexpected Similarities." *Journal of Ecumenical Studies* 31 (winter-spring 1994): 76–93.

———. *Christian Theology: An Eschatological Approach,* vol. 2. Scottdale, Pa.: Herald Press, 1989.

———. *Following Jesus Christ in the World Today.* Elkhart, Ind.: Institute of Mennonite Studies, 1984.

———. "Kingdom of God." *Mennonite Encyclopedia* 5 (1990): 491.

———. *Self, Earth, and Society.* Downers Grove, Ill.: InterVarsity Press, 1997.

———. "Trinity, Ecology, and Panentheism." *Christian Scholar's Review* 28 (fall 1997): 74–98.

Fox, Matthew. *The Coming of the Cosmic Christ.* New York: Harper & Row, 1988.

————. *Original Blessing.* Santa Fe: Bear, 1983.

————. *A Spirituality Named Compassion.* New York: Harper & Row, 1990.

Freedman, Ronald. "Norms for Family Size in Underdeveloped Areas." In *Population, Environment, and Social Organization: Current Issues in Human Ecology,* edited by Michael Micklin, 171–195. Hillsdale, Ill.: Dryden Press, 1973.

French, William C. "Ecological Degradation as the Judgment of God." *Christian Century* 110 (6–13 January 1993): 22–23.

Fretz, J. Winfield. "Farming among the Mennonites in North America." *Mennonite Encyclopedia* 2 (1956): 307–309.

————. *The Waterloo Mennonites: A Community in Paradox.* Waterloo, Ontario: Wilfrid University Press, 1989.

Friedmann, Robert. *The Theology of Anabaptism.* Scottdale, Pa.: Herald Press, 1973.

Friesen, Delores Histand. *Living More with Less Study / Action Guide.* Scottdale, Pa.: Herald Press, 1981.

Funk, Robert, and Roy Hoover. *The Five Gospels.* New York: Macmillan, 1993.

George, K. M. "Toward a Eucharistic Ecology: An Orthodox Perspective." *Reformed Journal* 40 (April 1990): 17–22.

Gex, Ken. "Has Awareness Brought Lifestyle Changes?" *MCC Contact* 15 (August 1991): 4–5.

Giglio, Virginia. *Southern Cheyenne Women's Songs.* Norman: University of Oklahoma Press, 1994.

Glick, Velda. "Church Called to More Leadership on Land Use Issues." *From Swords to Plowshares,* June 1979.

Good, Merle. "The Amish: Plain—but Not Ordinary—People." *Christianity Today* 34 (22 October 1990): 28.

Goode, William J. *Religion among the Primitives.* Glencoe, N.Y.: Free Press, 1951.

Gore, Al. *Earth in the Balance.* New York: Houghton Mifflin, 1992.

Gottwald, Norman. *The Hebrew Bible: A Socio-literary Introduction.* Philadelphia: Fortress, 1985.

Goudzwaard, Bob, and Harry de Lange. *Beyond Poverty and Affluence: Toward an Economy of Care.* Grand Rapids, Mich.: Eerdmans, 1995.

Goulet, Denis. *Development Ethics: A Guide to Theory and Practice.* New York: Apex Press, 1995.

Granberg-Michaelson, Wesley. *Tending the Garden: Essays on the Gospel and the Earth.* Grand Rapids, Mich.: Eerdmans, 1987.

————. *A Worldly Spirituality: The Call to Redeem Life on Earth.* San Francisco: Harper & Row, 1984.

Grebel, Conrad. "Letter to Müntzer, 1524." In *Anabaptism in Outline*, edited by Walter Klaassen, 191–192. Scottdale, Pa.: Herald Press, 1981.

Greenberg, Russell, and Jamie Reaser. *Bring Back the Birds.* Mechanicsburg, Pa.: Stackpole Books, 1995.

Gregorios, Paulos Mar. "New Testament Foundations for Understanding the Creation." In *Liberating Life: Contemporary Approaches to Ecological Theology*, edited by Charles Birch, William Eakin, and Jay B. McDaniel, 37–45. Maryknoll, N.Y.: Orbis Books, 1990.

Grove, Stanley. Electronic mail letter. Goshen, Ind., 5 August 1996.

Habgood, John. "A Sacramental Approach to Environmental Issues." In *Liberating Life: Contemporary Approaches to Ecological Theology*, edited by Charles Birch, William Eakin, and Jay B. McDaniel, 47–53. Maryknoll, N.Y.: Orbis Books, 1990.

Handrich Schlabach, Joetta. *Extending the Table: A World Community Cookbook.* Scottdale, Pa.: Herald Press, 1991.

Harden, Blaine. "Nuclear Reactions." *Washington Post Magazine*, 5 May 1996, 12–29.

———. *A River Lost: The Life and Death of the Columbia.* New York: W. W. Norton, 1966.

Hardin, Garret. "The Tragedy of the Commons." *Science* 162 (12 December 1968): 1243–1248.

Harris, Murray J. *Colossians and Philemon: Exegetical Guide to the Greek New Testament.* Grand Rapids, Mich.: Eerdmans, 1991.

Hart, John. *The Spirit of the Earth: A Theology of the Land.* New York: Paulist Press, 1984.

Hart, Julie. Electronic mail letter. North Newton, Kans., 18 July 1996.

Hauser, Philip M. *The Population Dilemma.* Englewood Cliffs, N.J.: Prentice-Hall, 1963.

Hawken, Paul. *The Ecology of Commerce: A Declaration of Sustainability.* New York: Harper Collins, 1993.

Hefner, Philip J. "The Sacramental Paradigm of Nature." *Currents in Theology and Mission* 20 (June 1993): 197–200.

Helwig, David Gordon. "One Step from an Old Dance." In *The Second Century Anthology of Verse*, edited by Roberta A. Charlesworth, 85. Toronto: Oxford University Press, 1969.

Henderson, Hazel. *The Politics of the Solar Age: Alternatives to Economics.* Garden City, N.Y.: Anchor Books, 1981.

Henry, Jules. *Culture against Man.* New York: Random House, 1963.

Hessel, Dieter T., ed. *Preaching, Ecology, and Justice.* Philadelphia: Geneva Press, 1985.

Hiebert, Ted. *The Yahwist Landscape: Nature and Religion in Early Israel.* New York: Oxford University Press, 1997.

Hirschfelder, Arlene, and Paulette Molin. *The Encyclopedia of Native American Religions.* New York: Facts on File, 1992.

Hoig, Stan. *The Peace Chiefs of the Cheyennes.* Norman, Okla.: University of Oklahoma Press, 1980.

Horsch, James E., ed. *Mennonite Yearbook and Directory, 1995.* Scottdale, Pa.: Mennonite Publishing House, 1995.

Hostetler, J. J. *Mennonite Business and Professional People's Directory, 1978.* Goshen, Ind.: MIBA, 1978.

Hostetler, John A. *Amish Society,* 4th ed. Baltimore: Johns Hopkins University Press, 1993.

Hulteen, Bob, and Jim Wallis, eds. *Who Is My Neighbor? Economics as If Values Matter.* Washington, D.C.: Sojourners, 1994.

Hymnal: A Worship Book. Scottdale, Pa.: Mennonite Publishing House, 1992.

Janzen Longacre, Doris. *Living More with Less.* Scottdale, Pa.: Herald Press, 1980.

———. *The More with Less Cookbook.* Scottdale, Pa.: Herald Press, 1976.

Jegen, Mary Evelyn, and Bruno V. Manno. *The Earth Is the Lord's: Essays in Stewardship.* New York: Paulist Press, 1978.

Joint Mennonite Environmental Task Force. *Report to Assembly.* Akron, Pa., 1995.

Kanagy, Conrad, and Donald B. Kraybill. "The Rise of Entrepreneurship in Two Old Order Communities." *Mennonite Quarterly Review* 70, no. 3 (1996): 263–279.

Kates, Robert W. "Population, Technology, and the Human Environment: A Thread through Time." *Daedalus* 125 (summer 1996): 43–71.

———. "Sustaining Life on Earth." *Scientific American* 271 (October 1994): 114–122.

Kauffman, J. Howard, and Leo Driedger. *Mennonite Mosaic.* Scottdale, Pa.: Herald Press, 1991.

Kauffman, J. Howard, and Leland Harder. *Anabaptists Four Centuries Later.* Scottdale, Pa.: Herald Press, 1975.

Kauffman, Milo. *Stewards of God.* Scottdale, Pa.: Herald Press, 1975.

Kearns, Laurel. "Saving the Creation: Christian Environmentalism in the United States." *Sociology of Religion* 57 (spring 1996): 55–69.

Kehm, George H. "Priest of Creation." *Horizons in Biblical Theology* 14 (December 1992): 129–142.

Kennedy, Margrit. *Interest and Inflation Free Money: Creating an Exchange Medium That Works for Everybody and Protects the Earth,* rev. ed. Philadelphia: New Society Publishers, 1995.

Kennell, Roger. Interview by the author. Roanoke, Ill., 6 August 1996.

"Kickapoo Valley Leads Assault on Flight Plan." *Wisconsin State Journal*, 8 October 1995.

Kirk, Janice E., and Donald R. Kirk. *Cherish the Earth: The Environment and Scripture*. Scottdale, Pa.: Herald Press, 1993.

Kittredge, William. "Home Landscape." In *Landscape in America*, edited by George F. Thompson, 143–153. Austin: University of Texas Press, 1995.

Klaassen, Walter. *Anabaptism: Neither Catholic nor Protestant*. Waterloo, Ont.: Conrad Press, 1981.

———. *Anabaptism in Outline*. Scottdale, Pa.: Herald Press, 1981.

———. "'Gelassenheit' and Creation." *Conrad Grebel Review* 9 (winter 1991): 23–36.

Kollmorgen, Walter. *Culture of a Contemporary Rural Community: The Old Order Amish of Lancaster, Pennsylvania*. Rural Life Studies Bulletin 4. Washington, D.C.: U.S. Department of Agriculture, 1942.

Korten, David C. *When Corporations Rule the World*. West Hartford, Conn.: Kumarian Press, 1995.

Krahn, Cornelius. "Business among the Mennonites in Russia." *Mennonite Encyclopedia* 1 (1956): 484–485.

Kraus, Norman. *God Our Savior*. Scottdale, Pa.: Herald Press, 1991.

Kraybill, Donald, and Stephen Nolt. *Amish Enterprises*. Baltimore: Johns Hopkins University Press, 1995.

Kritzinger, J. J. "Mission and the Liberation of Creation: A Critical Dialogue with M. L. Daniel." *Missionalia*. 20 (August 1992): 99–115.

Kropf, Marlene. "Let All Things Their Creator Bless." *Mennonite* 110 (27 June 1995): 10–11.

Krueger, Frederick. *Christian Ecology*. San Francisco: NACCE, 1988.

———. "The Effects of Television Viewing on Children and the Environment." *Green Cross* 2 (spring 1996): 18–21.

Kysar, Robert. *John the Maverick Gospel*. Atlanta: John Knox Press, 1976.

Land, Richard D., and Louis A. Moore, eds. *The Earth Is the Lord's: Christians and the Environment*. Nashville: Broadman Press, 1992.

Langin, Bernd. *Plain and Amish*. Scottdale, Pa.: Herald Press, 1994.

Lantz, Marvin. Interview by the author. Telephone, Archbold, Ohio, 30 July 1996.

Leonard, Bill. "From Smashed Pallets to Construction Blocks." *Des Moines Register*, 18 August 1996.

Leopold, Aldo. *A Sand County Almanac, with Other Essays from Round River*. New York: Oxford University Press, 1996.

Lestz, Gerald S. *Lancaster County's Firsts and Bests*. Lancaster, Pa.: John Baer's Sons, 1989.

Levinson, Marc. "Watching the Tide Rise." *Newsweek*, 11 October 1994, 47.

Livernash, Robert, and Eric Rodenburg. "Population, Change, Resources, and the Environment." *Population Bulletin* 53 (March 1998): 2–40.

Loewen, Arvid. Interview by the author. Telephone, Winnipeg, Manitoba, 25 July 1996.

Loewen, Howard John. *One Lord, One Church, One Hope, and One God.* Elkhart, Ind.: Institute of Mennonite Studies, 1985.

Loewen, Royden. *Family, Church, and Market: A Mennonite Community in the Old and New Worlds, 1850–1930.* Toronto: University of Toronto Press, 1993.

Longhurst, John. "Blue Boxes, Jobs: A Monument to Dave Hubert's Work with MCC Canada." *MCC News Service*, 9 February 1996, 1–2.

———. "Edmonton Recycling Society: A Model for Canada." *MCC News Service*, 9 February 1996, 2–3.

———. "'What Will Be the Costs?' Chemical Spraying Jeopardizes MCC Job Creation Efforts in Ontario." *MCC Contact*, August 1991, 3–4.

Lovelock, James. *The Ages of Gaia.* New York: W. W. Norton, 1988.

———. *Gaia: A New Look at Life on Earth.* New York: Oxford University Press, 1987.

Lowdermilk, C. W. *Conquest of the Land through Seven Thousand Years.* Bulletin 99. Washington, D.C.: U.S. Department of Agriculture, 1948.

Lucky, Robert. "What Technology Alone Cannot Do." *Scientific American* 273 (September 1995): 204–205.

Manahan, Ronald. "Christ as the Second Adam." In *The Environment and the Christian: What Does the New Testament Say about the Environment?* edited by Calvin B. DeWitt, 45–56. Grand Rapids: Baker Book House, 1991.

Margulis, Lynn, and Dorion Sasgan. *Microcosmos.* New York: McGraw-Hill, 1990.

Mazur, Laurie Ann, ed. *Beyond the Numbers: A Reader on Population, Consumption, and the Environment.* Washington, D.C.: Island Press, 1994.

McCarthy, Coleman. "The Noah Movement." *Washington Post*, 10 February 1996.

McClendon, James William, Jr. *Systematic Theology: Ethics*, vol. 1. Nashville: Abingdon, 1991.

McDaniel, Jay. *Earth, Sky, Gods, and Mortals.* Mystic, Conn.: Twenty-third, 1990.

McDonough, William, and Michael Braungart. "The Next Industrial Revolution." *Atlantic Monthly* 282 (October 1998): 82–92.

McFague, Sallie. *The Body of God.* Minneapolis: Fortress, 1993.

———. *Models of God.* Philadelphia: Fortress, 1987.

McKibben, Bill. "A Special Moment in History." *Atlantic Monthly* 281 (May 1998): 55–78.

McNamara, Robert S. *Africa's Development Crisis: Agricultural Stagnation, Population Explosion, and Environmental Degradation.* Washington, D.C.: Global Coalition for Africa, Institute for African Affairs. Typescript, 1991.

Meadows, Donella, Dennis L. Meadows, Joergen Randers, et al., eds. *The Limits to Growth.* New York: Universe Books, 1972.

"The Meditation of Marcus Aurelius." In *Encylopedia Britannica,* 266. Chicago: Encyclopedia Britannica, 1952.

Mennonite Encyclopedia. Scottdale, Pa.: Mennonite Publishing House.

"Mennonite Environmental Stewardship Census." Akron, Pa.: Joint Environmental Task Force, 1995.

The Mennonite Hymnary. Newton, Kans.: Mennonite Press, 1940.

Meyer, Art. *Christianity and the Environment: A Collection of Writings.* Occasional Paper No. 13. Akron, Pa.: Mennonite Central Committee, U.S. Peace Section Office, 1990.

Meyer, Art, and Jocele Meyer. *Earthkeepers: Environmental Perspectives on Hunger, Poverty, and Injustice.* Scottdale, Pa.: Herald Press, 1991.

Micklin, Michael. *Population, Environment, and Social Organization: Current Issues in Human Ecology.* Hillsdale, Ill.: Dryden Press, 1973.

Miller, Dennis. "Theology of Creation." *Mennonite Encyclopedia* 5 (1990): 210–211.

Miller, G. Tyler, Jr. *Living in the Environment: Principles, Connections, and Solutions,* 8th ed. Belmont, Calif.: Wadsworth Publishing, 1994.

Miller, Isaac. "Growing Up Anabaptist." *Christianity Today* 34 (22 October 1990): 27–28.

Miller, Levi. *Our People: The Amish and Mennonites of Ohio.* Scottdale, Pa.: Herald Press, 1983.

Miller, Marlin E. "America's Anabaptists: What They Believe." *Christianity Today* 34 (22 October 1990): 30–33.

Miller, Maynard. Interview by the author. Telephone, Holmesville, Ohio, 29 July 1996.

Minear, Paul. *Images of the Church in the New Testament.* Philadelphia: Westminster, 1960.

Minsky, Marvin. "Will Robots Inherit the Earth?" *Scientific American* 271 (October 1994): 108–113.

Moltmann, Juergen. *The Crucified God.* New York: Harper & Row, 1974.

———. *The Future of Creation.* Philadelphia: Fortress, 1979.

———. *The Trinity and the Kingdom.* New York: Harper, 1981.

————. *The Way of Jesus Christ*. San Francisco: Harper, 1990.

Nafziger, Estel Wayne. "Economics." *Mennonite Encyclopedia* 5 (1990): 255–256.

Nafziger, Joel. Electronic mail letter. Hopedale, Ill., 31 July 1996.

Napier, Ted L., and David G. Sommers. "Farm Production Systems of Mennonite and Non-Mennonite Land Owner-Operators in Ohio." *Journal of Soil and Water Conservation* 51 (January 1996): 71–76.

Neff, Robert W. "The Biblical Basis for Political Advocacy." *Brethren Life and Thought* 32 (autumn 1987): 201–208.

Negroponte, Nicholas. *Being Digital*. New York: Knopf, 1995.

Nitecki, Matthew, ed. *Evolutionary Progress*. Chicago: University of Chicago Press, 1988.

O'Brien, Sharon. *American Indian Tribal Governments*. Norman: University of Oklahoma Press, 1989.

O'Connor, Martin. *Is Capitalism Sustainable?* New York: Guilford Press, 1994.

Ortman, David. "More Ferrets and More Friends." *Forum*, April 1980, 9–10.

————. "What Good Is a Church without a Habitable Planet to Put It On?" *Source* 13 (December 1989): 1–2.

Penner, Larry. "Conference Aims to Create Green Theology." *Mennonite* 110 (14 March 1995): 13–14.

Peters, John. "Old Order Mennonite Economics." In *Anabaptist/Mennonite Faith and Economics*, edited by Calvin Redekop, Vic Krahn, and Samuel J. Steiner, 153–175. Lanham, Md.: University Press of America, 1994.

Petter, Rodolphe. *Cheyenne Spiritual Songs*. Lame Deer, Okla., 1942.

Piel, Jonathan, ed. *Managing Planet Earth: Readings from Scientific American Magazine*. New York: W. H. Freeman, 1990.

Platt, Anne E. "Confronting Infectious Diseases." In *State of the World*, edited by Lester Brown. New York: W. W. Norton, 1996.

————. "Dying Seas." *Worldwatch*, January/February 1995, 10–19.

Platt, LaVonne Godwin. *Hope for the Family Farm: Trust God and Care for the Family Farm*. Newton, Kans.: Faith & Life Press, 1987.

Polanyi, Karl. *The Great Transformation*. Boston: Beacon Press, 1944.

Postel, Sandra. "Carrying Capacity: Earth's Bottom Line." In *State of the World, 1994*, 3–21. New York: W. W. Norton, 1994.

————. "Protecting Forests from Air Pollution and Acid Rain." In *The State of the World*, edited by Lester Brown, 97–123. New York: W. W. Norton, 1985.

Price, Tom. "'What Vision Will Preserve Anabaptist Identity in the 21st Century?' Scholars Ask." *Gospel Herald*, 25 October 1994, 10–11.

Quinn, Frederick. *To Heal the Earth: A Theology of Ecology*. Nashville: Upper Room Books, 1994.

Raffensperger, Carolyn. Interview by the author. Telephone, 6 February 1996.

Ratzlaff Epp, Heinrich, as quoted in Juanita Thigpen, "The Savior of the Green Hell." *Nature Conservancy* 46 (May 1996): 12.

Redekop, Calvin. "Business." *Mennonite Encyclopedia* 5 (1991): 113–115.

―――. "Economic Developments among the United States Mennonite Brethren." In *Bridging Troubled Waters: The Mennonite Brethren at Mid-Twentieth Century*, edited by Paul Toews, 117–135. Winnipeg: Kindred Press, 1995.

―――. Electronic mail letter. Harrisonburg, Va., 30 March 1996.

―――. *Mennonite Society*. Baltimore: Johns Hopkins University Press, 1989.

―――. "Mennonites, Creation, and Work." *Christian Scholar's Review* 22 (June 1993): 348–366.

―――. "Mennonites, Work, and Economics." In *Anabaptist / Mennonite Faith and Economics*, edited by Calvin Redekop, Vic Krahn, and Samuel J. Steiner, 279–302. Lanham, Md.: University Press of America, 1994.

―――. *The Old Colony Mennonites*. Baltimore: Johns Hopkins University Press, 1969.

―――. "Toward a Mennonite Theology and Ethic of Creation." *Mennonite Quarterly Review* 60 (July 1986): 387–403.

―――. "Why a True-Blue Nonresistant Christian Won't Waste Natural Resources." *Festival Quarterly*, August-October 1977, 14–15.

Redekop, Calvin, Stephen C. Ainlay, and Robert Siemens. *Mennonite Entrepreneurs*. Baltimore: Johns Hopkins University Press, 1995.

Redekop, Calvin, Vic Krahn, and Samuel J. Steiner, eds. *Anabaptist / Mennonite Faith and Economics*. Lanham, Md.: University Press of America, 1994.

Redekop, Calvin, and Susan Shantz. "Mennonite Farm Practices and the Environment." Unpublished research report sponsored by MCC, 1988.

Redekop, Calvin, and Wilmar Stahl. "The Impact on the Environment of the Evangelization of the Native Tribes in the Paraguayan Chaco." *Evangelical Review of Theology* 17 (April 1993): 269–283.

Regehr, Ted. "The Economic Transformation of Canadian Mennonite Brethren." In *Bridging Troubled Waters: The Mennonite Brethren at Mid-Twentieth Century*, edited by Paul Toews, 97–115. Winnipeg: Kindred Press, 1995.

―――. "Mennonites and Entrepreneurship in Canada." In *Anabaptist / Mennonite Faith and Economics*, edited by Calvin Redekop, Vic Krahn, and Samuel J. Steiner, 111–125. Lanham, Md.: University Press of America, 1994.

Reiheld, Amelia T. "Buggy Travel Puts Amish in Slow Lane." *Homes County Hub*, June 1994, 11.

―――. "Living by Soil Is Living under the Hand of the Creator." In *The Amish: A Culture, a Religion, a Way of Life*. Special section of the *Homes County Hub*, June 1992.

Reimer, Margaret Loewen. "'How Does Peace Theology Relate to Other Faiths?' Conference Asks." *Gospel Herald*, 26 July 1994, 10.

Reumann, John. *Creation and New Creation*. Minneapolis: Augsburg, 1973.

Rhodes, Arnold B. *The Mighty Acts of God*. Richmond: CLC Press, 1964.

Ricard, François. "This Hippie Went to Market." *Saturday Night*, October 1994, 41–42.

Rich, Bruce. *Mortgaging the Earth: The World Bank, Environmental Impoverishment, and the Crisis of Development*. Boston: Beacon Press, 1994.

Rich, Jim. "Hope for the Planet." *Mennonite* 107 (23 June 1992): 271–272.

Rifkin, Jeremy. *Declaration of a Heretic*. Boston: Routledge & Kegan Paul, 1984.

Rimbach, James A. "All Creation Groans: Theology/Ecology in St. Paul." *Asia Journal of Theology* 1 (October 1987): 379–391.

Robertson, James. *The Sane Alternative*. St. Paul: River Basin Press, 1979.

Rolston, Holmes, III. "Does Nature Need to Be Redeemed?" *Horizons in Biblical Theology* 14 (December 1992): 143–172.

Rosner, Rhonda. "Produce Farming Is Venture for 16 Amish Families." In *The Amish: A Culture, a Religion, a Way of Life*. Special section of the *Homes County Hub*, June 1992.

Ruether, Rosemary. *Gaia and God*. San Fransisco: Harper, 1992.

Ruth, John L. "America's Anabaptists: Who They Are." *Christianity Today* 34 (22 October 1990): 25–29.

Saint John of Damascus. *On the Divine Images*. Crestwood, N.Y.: St. Vladimir's Seminary Press, 1980.

Santmire, II. Paul. *The Travail of Nature: The Ambiguous Ecological Promise of Christian Theology*. Philadelphia: Fortress, 1985.

Sauer, Carl Ortwin. *Collected Essays, 1963–1975*. Berkeley, Calif.: Turtle Island Foundation, 1981.

———. *Early Spanish Main*. Berkeley and Los Angeles: University of California Press, 1966.

———. *Land and Life*, edited by John Leighly. Berkeley: University of California Press, 1963.

———. *Northern Mists*. Berkeley: University of California Press, 1989.

Schlabach, Jotta Handrich. *Extending the Table*. Scottdale, Pa.: Herald Press, 1991.

Schlatter, John. Interview by the author. Telephone, Archbold, Ohio, 30 July 1986.

Schmidt, Melvin D. *God So Loved the World: Telling the Salvation Story as if the Earth Mattered*. Alexandria, Va.: Empire Video, 1995

Schmookler, Andrew Bard. *Fool's Gold: The Fate of Values in a World of Goods*. New York: Harper, 1993.

Schnaiberg, Allan. *The Environment: From Surplus to Scarcity.* New York: Oxford University Press, 1980.

Schreiber, William. *Our Amish Neighbors.* Chicago: University of Chicago Press, 1962.

Schrock-Shenk, David. "Disasters Are Not Always Natural: A Reflection on Ways to Prevent Conditions That Lead to Natural Disasters." *Mennonite* 107 (23 June 1992): 277.

————. Electronic mail letter. Akron, Pa., 22 July 1996.

Schumacher, E. F. *Small Is Beautiful: Economics as If People Mattered.* New York: Harper & Row, 1973.

Senner, Stan. Interview by the author, 16 February 1996.

————. "Making a Difference." In *Bring Back the Birds,* edited by Russell Greenberg and Jamie Reaser, 117–119. Mechanicsburg, Pa.: Stackpole Books, 1995.

Sensenig, Pearl, ed. *Jottings* 14 (May 1992): 1–5.

Serageldin, Ismail. "Ethics and Spiritual Values and Promotion of Environmentally Sustainable Development." *Ethics,* winter 1996, 3–4.

Sheldon, Joseph. Electronic mail letter. Grantham, Pa., 23 July 1996.

Shelly, Karl S. "On the Road to Jairus' House, Take Seriously Advocacy for the Poor." *Mennonite* 111 (25 June 1996): 12–13.

Sherrard, Philip. "Sacred Cosmology and the Ecological Crisis." *Green Cross,* fall 1996, 8–14.

Simon, Julian, and Herman Kahn, eds. *The Resource Earth: A Response to Global 2000.* Oxford: Basil Blackwell Publishers, 1984.

Simons, Menno. "The Blasphemy of John of Leiden." In *The Complete Writings of Menno Simons,* translated by Leonard Verduin, edited by J. C. Wenger, 31–50. Scottdale, Pa.: Herald Press, 1956.

————. "The Incarnation of Our Lord." In *The Complete Writings of Menno Simons,* translated by Leonard Verduin, edited by J. C. Wenger, 783–834. Scottdale, Pa.: Herald Press, 1956.

————. "Epistle to Martin Micron." In *The Complete Writings of Menno Simons,* translated by Leonard Verduin, edited by J. C. Wenger, 915–943. Scottdale, Pa.: Herald Press, 1956.

————. "Instructions on Excommunication." In *The Complete Writings of Menno Simons,* translated by Leonard Verduin, edited by J. C. Wenger, 959–998. Scottdale, Pa.: Herald Press, 1956.

————. "Reply to Gellius Faber." In *The Complete Writings of Menno Simons,* translated by Leonard Verduin, edited by J. C. Wenger, 623–781. Scottdale, Pa.: Herald Press, 1956.

————. "The Spiritual Resurrection." In *The Complete Writings of Menno Simons,* translated by Leonard Verduin, edited by J. C. Wenger, 51–62. Scottdale, Pa.: Herald Press, 1956.

Smucker, Art. Electronic mail letter. Goshen, Ind., 13 August 1996.

Smucker, Donovan. "Gelassenheit, Entrepreneurs, and Remnants: Socio-economic Models among the Mennonites." In *Kingdom, Cross, and Community*, edited by J. Richard Burkholder and Calvin Redekop, 219–241. Scottdale, Pa.: Herald Press, 1975.

Solow, Robert M. "Sustainability: An Economist's Perspective." In *Economics of the Environment: Selected Readings*, edited by Robert Dorfman and Nancy S. Dorfman, 180–181. New York: W. W. Norton, 1993.

Solzhenitsyn, Alexander. *From under the Rubble: Repentance and Self-limitation in the Life of Nations*. Boston: Little, Brown, 1974.

Sommer, Willis J. "Emanuel E. Mullet." In *Entrepreneurs in the Faith Community: Profiles of Mennonites in Business*, edited by Calvin W. Redekop and Benjamin W. Redekop, 98–122. Scottdale, Pa.: Herald Press, 1996.

Stoltzfus, Victor. "Amish Agriculture: Alternative Strategies for Economic Survival of Community Life." *Rural Sociology* 38 (1973): 196–205.

Stutzman, Clarence. Interview by the author. Telephone, Holmesville, Ohio, 29 July 1996.

Swimme, Brian, and Thomas Berry. *The Universal Story*. San Francisco: Harper, 1992.

"Tending God's Garden: Evangelical Group Embraces Environment." *Washington Post*, 17 February 1996.

Testa, Randy-Michael. *After the Fire: The Destruction of the Lancaster County Amish*. Hanover, N.H.: University Press of New England, 1992.

Thigpen, Juanita. "The Savior of the Green Hell." *Nature Conservancy Magazine* 46, no. 5 (1996): 10–15.

Thomas, Carolyn. "Romans 8: A Challenge to Christian Involvement." *Trinity Seminary Review* 10 (spring 1988): 31–37.

Thomas, Everett. Interview by the author, 13 July 1996.

Thompson, George F. *Landscape in America*. Austin: University of Texas Press, 1995.

Thompson, Marianne Meyer. *The Humanity of Jesus in the Fourth Gospel*. Philadelphia: Fortress, 1988.

Thoren, Theodore R., and Richard F. Warner. *The Truth in Money Book*, 4th ed. Chagrin Falls, Ohio: Truth in Money, 1994.

Todaro, Michael P. *Economic Development*, 5th ed. New York: Longman, 1994.

Toews, Paul. *Bridging Troubled Waters: The Mennonite Brethren at Mid-Twentieth Century*. Winnipeg: Kindred Press, 1995.

Toynbee, Arnold J. *A Study of History.* New York: Oxford University Press, 1947.

Traub, Bruce H. Electronic mail letter. Fresno, Calif., 26 July 1996.

Tsese-Ma'heone-Nemeototse. *Cheyenne Spiritual Songs,* hymn 95. Newton, Kans.: Faith & Life Press, 1982.

Turner, Jonathan H. *The Structure of Sociological Theory,* 3d ed. Homewood, Ill.: Dorsey Press, 1982.

United Nations Development Program (UNDP). *Human Development Report.* New York: Oxford University Press, 1992.

Van Dyke, Fred, David Mahan, Joseph Sheldon, and Raymond Brand, eds. *Redeeming Creation: The Biblical Basis for Environmental Stewardship.* Downers Grove, Ill.: InterVarsity Press, 1996.

Van Leeuwen, Raymond C. "Christ's Resurrection and the Creation's Vindication." In *The Environment and the Christian: What Does the New Testament Say about the Environment?* edited by Calvin B. Dewitt, 57–71. Grand Rapids, Mich.: Baker Book House, 1991.

Vogt, Roy. "Mennonite Studies in Economics." *Journal of Mennonite Studies* 2 (1983): 64–78.

Wackernagel, Mathis, and William Rees. *Our Ecological Footprint: Reducing Human Impact on Earth.* Gabriola Island, B.C.: New Society Publishers, 1996.

Warren, Karen. "A Feminist Philosophical Perspective of Ecofeminist Spiritualities." In *Ecofeminism and the Sacred,* edited by Carol J. Adams, 119–132. New York: Continuum, 1993.

Wauzzinski, Robert A. "Technological Optimism." *Journal of the American Scientific Affiliation* 48 (September 1996): 144–151.

Weatherford, J. McIver. *Indian Givers: How the Indians of the Americas Transformed the World.* New York: Crown Publishers, 1988.

Weaver, J. Denny. *Becoming Anabaptist.* Scottdale, Pa.: Herald Press, 1987.

Whalen, Charles J. *The Anxious Society: Middle Class Insecurity and the Crisis of the American Dream,* 9–11. Report. Jerome Levy Economics Institute of Bard College, October 1995.

White, Lynn. "The Historical Roots of Our Ecologic Crisis." *Science* 155 (1967): 1203–1207.

Wiebe, Rudy. *The Temptations of Big Bear.* Toronto: McClelland & Stewart, 1973.

Wiebe-Powell, Wendell. "Reflections on the Creation Summit: Shaping an Anabaptist Theology for Living." Unpublished paper, Akron, Pa., 1995.

Wiener, Philip P., ed. *Dictionary of the History of Ideas,* vol. 2. New York: Charles Scribner's Sons, 1973.

Wilkinson, Loren. "Christ as Creator and Redeemer." In *The Environment and the Christian: What Does the New Testament Say about the Environment,* 25–44. Grand Rapids: Baker Book House, 1991.

————. "How Christian Is the Green Agenda?" *Christianity Today* 37 (11 January 1993): 16–20.

Willard, Janet. "Amish Farm by Rules of Church and God." In *The Amish: A Culture, a Religion, a Way of Life.* Special section of the *Holmes County Hub* (June 1992).

Wissler, Clark. *The Relation of Nature to Man in Aboriginal America.* New York: Oxford University Press, 1926.

World Bank. *World Development Report.* New York: Oxford University Press, 1995.

World Commission on Environment and Development. *Our Common Future.* London: Oxford University Press, 1987.

World Resources Institute. *World Resources, 1994–1995.* New York: Oxford University Press, 1994.

World Watch Institute. *State of the World.* New York: W. W. Norton, 1991.

Wright, G. Ernest. *God Who Acts: Biblical Theology as Recital.* Studies in Biblical Theology, First Series, No. 8. London: SCM, 1952.

Wright, N. T. *The New Testament and the People of God.* Minneapolis: Fortress, 1992.

Wuthnow, Robert, ed. *Rethinking Materialism: Perspectives on the Spiritual Dimension of Economic Behavior.* Grand Rapids, Mich.: Eerdmans, 1995.

Wysham, Daphne. "Ten-to-One Against: Costing People's Lives for Climate Change." *Ecologist* 24 (November/December 1995): 204.

Yoder, John Howard. "Business among the Mennonites in France and Switzerland." *Mennonite Encyclopedia* 1 (1955): 481.

————. "Farming among the Mennonites in France." *Mennonite Encyclopedia* 2 (1956): 306–307.

Yoder, Larry. Electronic mail letter. Goshen, Ind., 20 August 1996.

Yoder, Michael L. "The Family Farmer: From Boom to Bust." *Church Herald* 43, no. 3 (1986): 5–7.

————. "Findings from the 1982 Mennonite Census." *Mennonite Quarterly Review* 59, no. 4 (1985): 307–349.

Yoder, Robert A. "Holy Disturbance." Unpublished poem received by electronic mail. Eureka, Ill., 10 April 1996.

Young, Norman. *Creator, Creation, and Faith.* Philadelphia: Westminster, 1976.

Zachary, G. Pascal. "A 'Green Economist' Warns Growth May Be Overrated." *Wall Street Journal,* 25 June 1996.

Zerbe, Gordon. "Ecology According to the New Testament." *Direction* 21 (fall 1992): 15–26.

———. "The Kingdom of God and Stewardship of Creation." In *The Environment and the Christian: What Does the New Testament Say about the Environment?* edited by Calvin B. DeWitt, 73–92. Grand Rapids, Mich.: Baker Book House, 1991.

Contributors

Heather Ann Ackley Bean (Ph.D. in theology, Claremont Graduate School) is an ordained minister of the Pacific Southwest Mennonite Conference. She has focused on women's studies and the ethical aspects of Mennonite environmental behavior. Her publications include "Eating God: Beyond a Cannibalizing Christology" in *Process Studies* (1992).

Kenton K. Brubaker (Ph.D. in biology, University of Ohio) is professor emeritus of biology at Eastern Mennonite University. He has taught environmental courses, practiced conservation, and researched and written on ecological issues for many years. He is the author of *La cellule: Structure et physiologie cellulaire* (1965).

Thomas Finger (Ph.D in theology, Claremont Graduate School) is a professor of theology at Eastern Mennonite Seminary. He has been involved in numerous interchurch organizations concerned with the environment and has produced many publications on the theology of creation, including *Self, Earth, and Society* (1997) and *Christian Theology: An Eschatological Approach* (1989).

James M. Harder (Ph.D. in economics and development, Notre Dame University) has served under the Mennonite Central Committee in Africa. Harder is chairman of the Economics Department at Bethel College. He is the author of "The Economics of War" (1991) and "Ethics and the World Trading System" (1995).

Karen Klassen Harder (Ph.D. in consumer economics, Purdue University) has served under the Mennonite Central Committee in Africa and is now chairwoman of the Mennonite Central Committee. She is also chairwoman of the Social Sciences Department at Bethel College in Kansas.

Lawrence Hart is a member of the Cheyenne nation and a traditional peace chief. He graduated from Bethel College and attended Mennonite Biblical Seminary in Chicago. He is executive director of the Cheyenne Cultural Center at Clinton, Oklahoma.

Theodore Hiebert (Ph.D. in classical Hebrew and Hebrew Scriptures, Harvard University) is an associate professor at McCormick Theological Seminary. He has written extensively on the influence of biblical thought on the environment; examples are *The Yahwist's Landscape: Nature and Religion in Early Israel* (1996) and "The Human Vocation: Origins and Transformations in Christian Traditions" (1999).

Carl Keener (Ph.D. in biology, North Carolina State University) is professor of biology (emeritus) at Pennsylvania State University and has written extensively on philosophical, theological, population, and biological issues in such journals as *Molecular Phylogenetic Evolution* and *Taxon*.

Walter Klaassen (Ph.D in history, Oxford University) is professor emeritus at Conrad Grebel College in Ontario. He is an expert on Anabaptist/Mennonite history and theology and has written and spoken widely on creation/environment issues. Two of his books are *Armageddon and the Peaceable Kingdom* (1999) and *Anabaptism: Neither Catholic nor Protestant* (1981).

David Kline, an Amish person who "sleeps in the room in which he was born," has spoken and written widely about the Amish view of environment and farms 120 acres of rolling land in Holmes County, Ohio. He writes a weekly column, "On Nature," for the Wooster, Ohio, *Daily Record.*

Calvin Redekop (Ph.D. in sociology and anthropology, University of Chicago) is professor emeritus at Conrad Grebel College. He has been involved in environmental conservation and has researched and written about Anabaptism and Mennonites in economics, culture, and the environment. His books include *Mennonite Society* (1989) and *Strangers Become Neighbors: Mennonite and Indigenous Relations in the Paraguayan Chaco* (1980).

Mel Schmidt has studied at Associated Mennonite Biblical Seminaries, has pastored at Mennonite congregations, is an environmental activist, and has conducted numerous seminars and produced videos promoting the preservation of the environment. He is a contributing editor and writer for *Builder*, a monthly magazine.

Dorothy Jean Weaver (Ph.D. in New Testament, Union Theological Seminary, Richmond, Virginia) is associate professor of New Testament at Eastern Mennonite University and has researched the New Testament on a variety of social and environmental issues. She wrote *Matthew's Missionary Discourse: A Literary Critical Analysis* (1990) and "Power and Powerlessness: Matthew's Use of Irony in the Portrayal of Political Leaders" (1996).

Michael L. Yoder (Ph.D. in sociology and statistics, University of Wisconsin–Madison) is a professor of sociology at Northwestern College, Iowa. He has studied and written on Mennonite demography and rural life. He is the author of "Transforming Society vs. Purifying the Church: Reformed, Amish and Mennonites as a Case Study" (1997) and "Coming to Terms with Marxism" (1980).

Index

technological revolution: in agriculture, 28; Anabaptist land stewardship affected by, 74–76; toolmaking affected by, 28

technology: appropriate, 84–85; belief in, 13; controlled by community, 197; critiqued, 14–16; effects of, 27; energy-producing, 82; and environmental degradation, 66; and ideology of growth, 229n. 63; inadequate, 13; optimum use of, 35; out of control, 37

theology, Anabaptist: and creation, 184; and dualistic nature, 184; focus on selfhood, 186; leaning toward human domination of earth, 185; problems with, 183–184; and reconciliation of earth, 193; restricted to nonviolence, 185; and view of world, 243n. 6; weak regarding creation, 184–185

theology, environmental: based on Jesus' love ethic, 201; derived from life in Christ, 201; marginal in Mennonite confessions, 239n. 2; weak in New Testament, 238n. 5

theology of creation: Anabaptist, 154–156; —, never fully developed, 169; demanding care of nature, 167; demanding trinitarian perspective, 169

Three Mile Island, 27

two-kingdoms theology, Anabaptist support of, 184–185

utilitarian philosophy, 52

vegetarianism: and care of creation, 192; mandated in creation, 196

violence against nature, 142; Anabaptist role in, 142–143; among Mennonites, 146; rejected in peace theology, 195–197

visions. *See* economic visions

voluntary service, Mennonites in, 83

waste as food, new ethic, 212

water: consumption of, 42; quality of, 42–43

wealth: and consumption, 188; trickle-down theory of, 18; unequal distribution of, 17

Western culture, one-directional causality, 52; utilitarian, 52

wind turbines, 82

women, status in childbearing, 227n. 31

world: defined by Anabaptists, 243n. 6; Anabaptist perception of, 246n. 1

World Bank, 4, 19

World Trade Organization, 4

World Watch Institute, 12

Yahwist: concern for care of creation, 117–118; concern for productivity, 115–117; view of creation, 113–116

zero population growth, 48, 56, 226n. 25

NAMES

Aquinas, Thomas, 47
Augustine, 146
Aurelius, Marcus, 213